Recent Progress in Integrated Photonics

Recent Progress in Integrated Photonics

Guest Editor

Xiaoyan Zhou

MDPI

Basel • Beijing • Wuhan • Barcelona • Belgrade • Novi Sad • Cluj • Manchester

Guest Editor
Xiaoyan Zhou
School of Precision
Instruments and
Opto-Electronics Engineering
Tianjin University
Tianjin
China

Editorial Office
MDPI AG
Grosspeteranlage 5
4052 Basel, Switzerland

This is a reprint of the Special Issue, published open access by the journal *Photonics* (ISSN 2304-6732), freely accessible at: https://www.mdpi.com/journal/photonics/special_issues/8V72Y537LQ.

For citation purposes, cite each article independently as indicated on the article page online and as indicated below:

Lastname, A.A.; Lastname, B.B. Article Title. *Journal Name* **Year**, *Volume Number*, Page Range.

ISBN 978-3-7258-6746-2 (Hbk)
ISBN 978-3-7258-6747-9 (PDF)
https://doi.org/10.3390/books978-3-7258-6747-9

Contents

About the Editor

Xiaoyan Zhou

Xiaoyan Zhou is currently an Associate Professor. She obtained her Bachelor and PhD degree from Tianjin University and has worked as a Postdoctoral Research and Assistant Professor at Niels Bohr Institute (with Prof. Peter Lodahl), University of Copenhagen, Denmark. Her research interests include quantum photonics, photonic computing, micro-/nano-opto-electro-mechanical systems, and hybrid integration on silicon platform. She has published many peer-reviewed journals and conference papers, including *Nature Nanotechnology, Science Advances, Optica, Laser and Photonics Reviews, Photonics Research,* and *Photonics Insights,* and has a h-index of 16 with a total citation rate >1000. She hosts a Young Researcher Project supported by the National Natural Science Foundation of China (NSFC). She is a member of OSA and assistant editor for *Frontiers in Photonics.* She is also a frequent reviewer for many journals, including *Light: Science & Applications, ACS Photonics, Optics Letters, IEEE Journal of Lightwave Technology,* etc.

photonics

MDPI

Article

Compact Nanolaser Relying on Bound States in the Continuum with Simultaneous Pump and Emission Enhancement

Yifei Ma, Jingyuan Ji, Xiaoyan Zhou * and Lin Zhang *

State Key Laboratory of Precision Measuring Technology and Instruments, Key Laboratory of Opto-Electronic Information Technology of Ministry of Education, Tianjin Key Laboratory of Integrated Opto-Electronics Technologies and Devices, School of Precision Instruments and Opto-Electronics Engineering, Tianjin University, Tianjin 300072, China; yifei_matju@163.com (Y.M.); jijingyuan_123@163.com (J.J.)

* Correspondence: xiaoyan_zhou@tju.edu.cn (X.Z.); lin_zhang@tju.edu.cn (L.Z.)

Abstract: Bound states in the continuum (BICs), characterized by high-Q modes, have demonstrated exceptional capabilities for enhancing light-matter interactions and, when combined with gain media, can enable compact lasers with low threshold power. However, conventional BIC lasers typically rely on the emitting light forming a BIC mode, leading to vertical emission, and often lack mechanisms to enhance pump efficiency. In this work, we propose a photonic crystal laser design that incorporates high-Q modes at both pump and emitting wavelengths. The pump light at 980 nm is designed to form a BIC state near the Γ-point, while the emitting light at 1550 nm is confined within a bandgap-defined cavity mode at the M-point, allowing efficient in-plane emission. This design leads to a compact footprint of 19.7 × 17.1 μm^2 and predicts a significant reduction in threshold power compared with a laser with a single resonance at the emission wavelength, providing a promising approach for developing compact on-chip lasers with significantly improved efficiency.

Keywords: bound states in the continuum; photonic crystals; bandstructure; on-chip laser

1. Introduction

Ultra-low threshold and compact-size lasers are very important in photonic integrated circuit design [1–4]. With increasing research effects on the miniaturization of lasers, the size of lasers has been reduced to the wavelength scale exploiting different interaction mechanisms [5–7]. Among various types of structures, planar photonic crystal (PhC) lasers consisting of periodic dielectric structures are a promising platform for constructing compact and high-quality (Q) factor lasers, where a wavelength-scale photonic mode can be formed through optical bandgap [5]. More recently, bound states in the continuum (BICs) have been demonstrated as a very efficient way to significantly boost the Q factors of planar PhCs [8–10]. The BIC state can be understood as an exception to the continuous modes in that its frequency exists inside the continuous spectrum, but it can stay confined without outward radiation [11,12]. Apart from exhibiting high-Q resonances, BICs featuring topological properties [13–15] can be exploited for the generation of vortex beams [16–18] and the formation of unidirectional radiation structures [19,20]. The topological feature of BICs also provides BIC PhC lasers robustness against structural disorder, when compared to conventional PhC lasers working in bandgap-confined modes [21,22].

Most BICs rely on periodic and symmetric structures to achieve a high Q-factor, which typically necessitates large PhC structures consisting of 10s to 100s of unit cells, thereby limiting the miniaturization of BIC lasers [16]. To mitigate the impact of truncation on the

Q-factor, PhC structures incorporating so-called super-BICs—formed by merging multiple BIC modes—have been proposed, enabling the maintenance of low thresholds in the finite-size BIC lasers [21,23]. Alternatively, BIC lasers that achieve three-dimensional light trapping were recently introduced, relying on PhC structures supporting a so-called mini-BIC mode, which combines BICs with a lateral PhC bandgap-confined cavity mode. This approach achieved a record-high Q-factor of 32,500 and single-mode lasing with a low threshold of 80 W/cm^2 in one study [24]. In another work, by enabling the trapping of both light and carriers in all three dimensions, the single-mode lasing threshold was further reduced to 52 W/cm^2 in a cavity size of just 5 × 5 unit cells (~2.5 × 2.5 μm^2) [1]. Despite these advancements, most BIC lasers rely on the emitting light forming the BIC mode, which results in surface emission and hinders integration with other on-chip photonic devices. Additionally, these lasers often lack mechanisms to enhance pump light, which, if properly exploited, could further improve light-matter interactions and provide an additional pathway to lower the lasing threshold [25–27].

In this paper, we design an AlGaAs PhC slab incorporating a gain medium of Er-doped Al_2O_3. This PhC structure supports high-Q modes at both the pump wavelength (980 nm) and the emitting wavelength (1550 nm). The 980 nm mode is a BIC at the Γ-point, while the 1550 nm mode is a PhC cavity mode within the bandgap at the M-point. The structure allows for simultaneous field enhancement of both the pump and emitting light, with calculated Q-factors of 3.9 × 10^2 and 2.3 × 10^3 within a compact size of 19.7 × 17.1 μm^2, respectively, and a 390-fold reduction in threshold power compared with a laser whose only resonant at the emission wavelength is predicted. Moreover, the emitting light is enhanced by the PhC cavity mode at the M-point, which can be easily extracted into in-plane photonic devices, facilitating its usage as an on-chip light source.

2. Unit-Cell Design and Bandstructure of the PhC Slab

In a four-energy level system, the threshold power is given by [28]:

$$P_{th} = \frac{h\nu_p A\alpha_s/(\Gamma_s\sigma_{em}N)}{\Gamma_p\sigma_{abs,p}\tau E(1-\alpha_s/(\Gamma_s\sigma_{em}N))} \tag{1}$$

where $\sigma_{abs,p}$ is the absorption cross-sectional area for the pump light, σ_{em} is the cross-sectional area for stimulated emission of the signal light, E is the intensity enhancement factor for the resonant pump, N is the dopant concentration, τ denotes the spontaneous emission lifetime, h is Planck constant, ν_p is the frequency of pump light, Γ_p and Γ_s are fractional intensities inside the microfiber for the pump and signal light, A is the cross-sectional area of the microfiber, and α_s is the loss coefficient. We note that α_s is related to the Q-factor of the resonant cavity in the signal mode by $Q_e = 2\pi n/\lambda\alpha_s$, where n is the refractive index and λ is the free space wavelength. When Q_e is large (α_s is small), $1-\alpha_s/(\Gamma_s\sigma_{em}N)$ is approximately 1, and P_{th} can be simplified to:

$$P_{th} = \frac{2\pi nh\nu_p A\alpha_s/(\Gamma_s\sigma_{em}N)}{\lambda Q_e\Gamma_p\sigma_{abs,p}\tau E} \tag{2}$$

in which case P_{th} is inversely proportional to Q_e. For the pump light, we note that the intensity build-up in a resonator compared to the input is related to its finesse (F) and Q factor as $B \propto F \propto Q_p$ [29], which should result in a threshold power of P_{th}/B compared to the non-resonant case. Therefore, the threshold power of a photonic cavity is simultaneously in resonance with the pump and emitting wavelength $P_{th} \propto 1/(Q_pQ_s)$.

Our PhC nanolaser is schematically shown in Figure 1, where the planar PhC is fabricated from a layer of $Al_xGa_{(1-x)}As$ on silicon oxide. AlGaAs stands as a compelling platform for C-band all-optical processing due to its balanced attributes of robust nonlinearity, low optical losses, and strong electro-optic effect [30,31]. While these properties support passive and modulatory components, complete photonic systems require integrated gain capabilities. The gain is provided by Er-doped aluminum, which is filled in the holes of the PhC structure. Incorporating Er^{3+} ions could enable the monolithic integration of lasers and amplifiers with existing AlGaAs photonic components, establishing an integrated semiconductor architecture for the generation, amplification, and processing of optical signals. Technically, this can be done through atomic layer deposition [32–34]. From the multi-energy model of erbium, the 980 nm wavelength light is typically used as the pump light to excite erbium, which, subsequently, emits at the wavelength of 1550 nm [35–37]. In $Al_xGa_{(1-x)}As$, x is set to 0.2 so that the material exhibits very low loss at both 980 nm and 1550 nm [38]. Moreover, the large refractive index contrast between AlGaAs and Al_2O_3 is desirable for forming a large bandgap in the photonic crystal, facilitating the construction of high-Q PhC cavity mode.

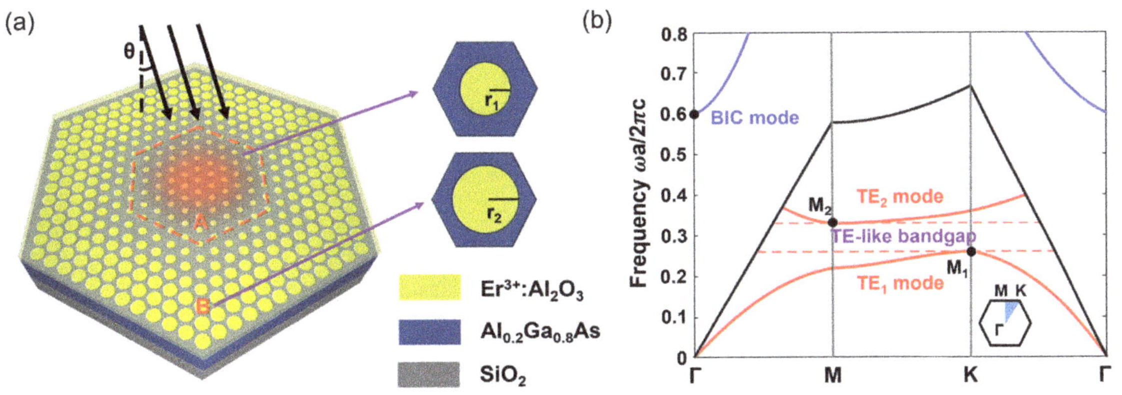

Figure 1. The working principle of the PhC nanolaser. (**a**) Schematic of the structure. Pump light (black) is incident in the vertical direction and forms BIC mode, while emitted light (red) forms a TE-like gap-confined cavity mode. (**b**) Schematic of the bandstructure, where BIC mode is formed at the pump frequency and TE-like gap-confined cavity mode is formed at the emitting frequency.

In order to form both BIC mode and PhC cavity mode for 980 nm and 1550 nm, as shown schematically in Figure 1a, different hole radii are chosen for regions A and B. The inset is a schematic representation of the 2D structure with different hole radii. The pump light is designed to incident into the structure in a vertical direction, forming a BIC mode at the Γ-point in region A, and the emitted light forms a cavity mode in the horizontal direction, where the radii in region B increase gradually to form a bandgap for the emitted light in the outer part. The bandstructure is shown schematically in Figure 1b. Within the light cone, the two red curves represent the two modes supported by the PhC, where the TE_1 mode has the lowest frequency and corresponds to the fundamental TE mode of the structure, and the TE_2 mode has the second highest frequency and corresponds to the second-order TE mode of the structure. There is a bandgap between the highest frequency M_1 of the TE_1 mode and the lowest frequency M_2 of the TE_2 mode, which does not allow the propagation of any light in this frequency range. Above the light cone, in the Γ direction, there are BIC states with infinitely high Q factors, which are based on the principle of destructive interference of the radiated light field and constructive interference of the light field within the material. The blue curve corresponds to the fundamental TE-BIC modes. As the wave vector is deflected in the vertical direction, the frequency of the mode changes and the Q factor decreases from infinity to finite values.

We set the thickness of the $Al_{0.2}Ga_{0.8}As$ layer to 220 nm, the lattice constant a = 564.1 nm, and the thickness of the alumina layer to 300 nm. We simulated the eigenfrequency variation of the hole radius from 235 nm to 265 nm. In order to form different energy bands within regions A and B, respectively, we choose different hole radii within the region A and B. As the hole radius increases, the effective refractive index of the structure within the cycle decreases, and the resonance wavelength of the BIC state changes. As shown in Figure 2a, the variation of the resonance wavelength with increasing radius is about −5.7 nm/nm, and the wavelength of the BIC state is 980.0 nm at a hole radius of 242.3 nm. Then, we solve for the wavelengths of the highest-frequency mode of the TE_1 energy band and the lowest-frequency mode of the TE_2 energy band around a = 242.3 nm. As shown in Figure 2b, the wavelength of light in the M_2 mode is 1550 nm at a radius of 242.3 nm. As the wavelength increases, the bandgap of the structure becomes narrower and the wavelengths of M_1 and M_2 are shifted to shorter wavelengths. The changes in their mode wavelengths with increasing radius are −7.7 nm/nm and −2.1 nm/nm, respectively. At a radius of 258.7 nm, the wavelengths of the M_1 and M_2 modes are 1516.3 nm and 1584.5 nm, respectively, and the 1550 nm wavelength is located in the middle of the bandgap, at which time the 1550 nm light cannot propagate in any direction. If we set the hole radius r_1 in region A to 242.3 nm, the structure can form a BIC state resonance cavity at the pump light wavelength of 980 nm and increase the hole radius r_2 in region B gradually to 258.7 nm, which can be used as a reflector to confine the 1550 nm emitting light generated by the stimulated radiation within the PhC cavity.

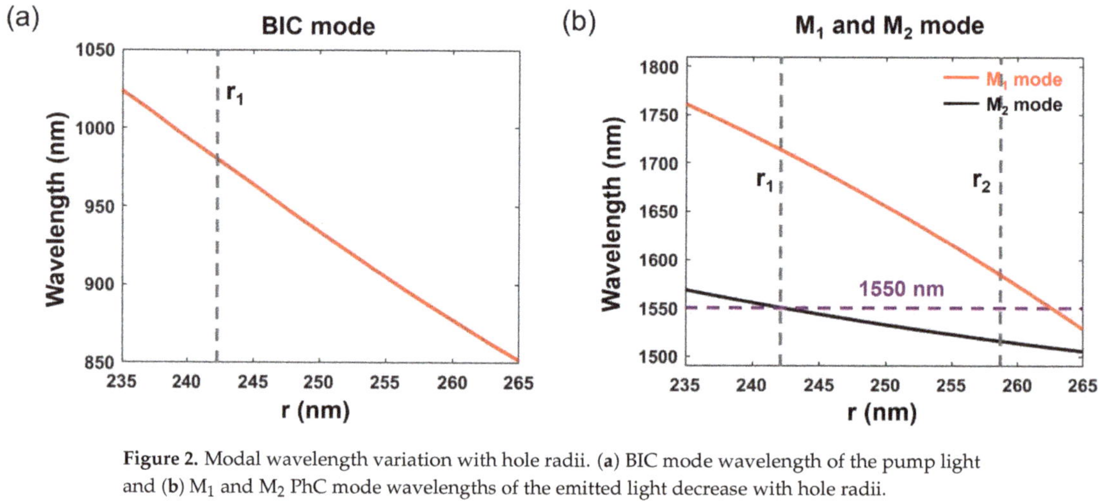

Figure 2. Modal wavelength variation with hole radii. (**a**) BIC mode wavelength of the pump light and (**b**) M_1 and M_2 PhC mode wavelengths of the emitted light decrease with hole radii.

We simulate the eigenmodes of the structure at different wave vectors for the structure at radii r_1 and r_2, respectively. In the structure with infinite periodic lattice at radius r_1 = 242.3 nm, the mode distribution of the lowest-frequency TE-BIC mode is shown in Figure 3a, where the plane is the x-y cross-section at the height of the center of the hole, and the mode has a Q factor of 1.5×10^7 and a wavelength of 980.0 nm. Calculating the eigenmodes of the structure in the M-direction, we obtain the mode field distribution of the M_2 mode in Figure 3b, which has a Q factor of 7.4×10^7 and a wavelength of 1550.0 nm. The mode overlap between the BIC mode and M_2 mode is 0.25. To further improve the overlap integral of the two modes, potentially one could seek for higher-order BIC mode as the pump light, or deploy alternative PhC materials with an even larger refractive-index contrast. The modes of the structure in the bandstructure are shown in Figure 3c, where the two lowest frequency modes existing below the light cone and the M_2 mode have a wavelength of 1550 nm. The TE-BIC fundamental mode is at the pump wavelength of 980 nm in the Γ direction; near the Γ direction, there are quasi-BIC states with finite Q and linewidth, and the Q factor of the modes corresponding to the blue line in the figure are all larger than 1.0×10^3. Under the condition of radius r_2 = 258.7 nm, the mode field distribution of TE-BIC mode is shown in Figure 3d. The resonant wavelength of the TE_2 mode in the M direction is 1516.3 nm, and its mode field distribution is shown in Figure 3e, which is similar to the mode field distribution in Figure 3a,b. The observed symmetry differences between the BIC mode and the PhC cavity mode stem from the distinct wavevector orientations at Γ and M points in the hexagonal lattice as Figure 1b. At the Γ point as Figure 3a,d, modes inherit the lattice's full hexagonal symmetry (C_6v), yielding radially symmetric profiles. By contrast, M-point modes as Figure 3b,e align with mirror-symmetry axes of the reduced Brillouin zone (C_2v symmetry), resulting in directional field confinement along the M vector. The corresponding bandstructure is shown in Figure 3f. Here the resonance wavelength of the TE_1 mode in the K-direction is 1584.5 nm, and the emitted light with wavelength 1550 nm moves from the originally guided-mode TE_2 to the middle of the bandgap, as shown by the purple dashed line, forming the reflector of the PhC cavity.

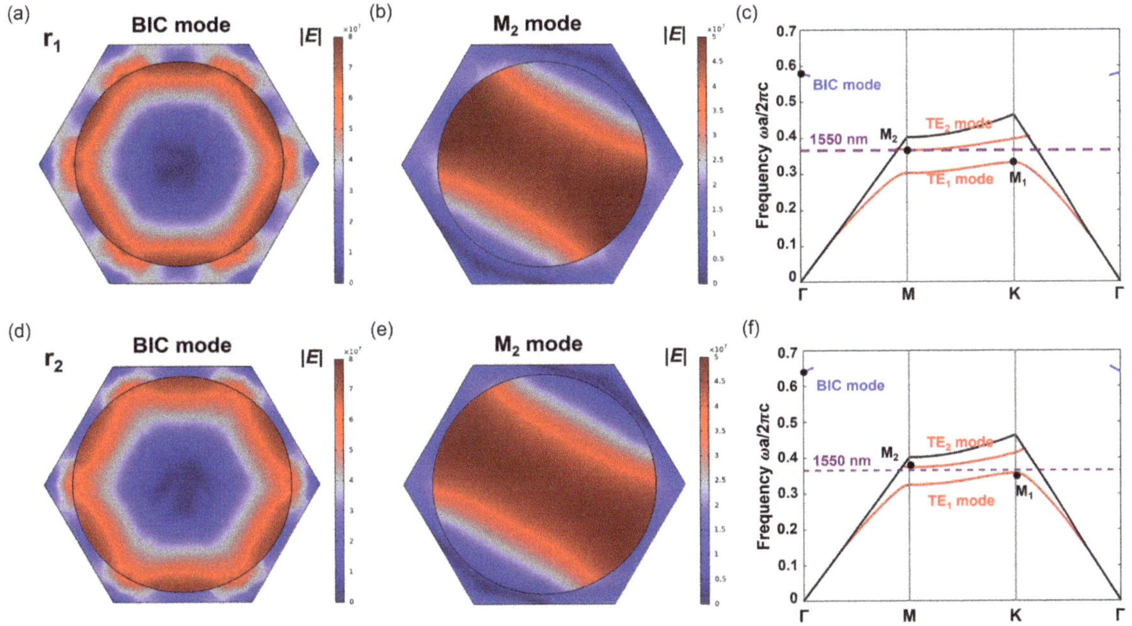

Figure 3. Eigenmodes of the PhC structure. (**a**) Mode field distribution of BIC modes at radius r_1. (**b**) Mode field distribution of M_2 modes at radius r_1. (**c**) Bandstructure of the structure at radius r_1. (**d**) Mode field distribution of BIC modes at radius r_2. (**e**) Mode field distribution of M_2 modes at radius r_2. (**f**) Bandstructure of the structure at radius r_2.

Theoretically, the BIC state has an infinite high Q and a vanished linewidth, but, in practice, the BIC state becomes a quasi-BIC state with finite high Q due to material absorption, dimensional effect, etc. As the angle of incidence deviates from the vertical direction, the symmetry of the PhC structure decreases with respect to it. The transmittance of the structure can be defined as $T = |t|^2$, where t is the amplitude of the transmitted wave. It can be seen from Figure 4a that the structure has an infinitely narrow resonance peak at the incidence angle θ = 0, and the resonance peak broadens as the incidence angle increases. As shown in Figure 4b, the BIC with infinite Q-factor changes into a high-Q quasi-BIC with increasing radiation loss as θ angle increases. Since the Q factor is infinite in the vertical incidence direction in the unit-cell simulation, it is necessary to slightly tilt the incidence of 980 nm light into the structure in Figure 3.

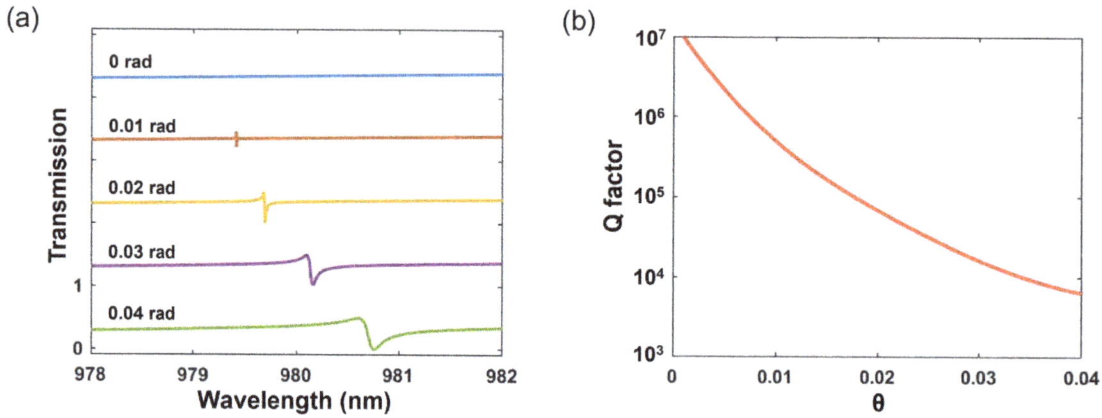

Figure 4. Evolution of (**a**) transmission and (**b**) Q factor of the BIC mode in the PhC structure with the angle of incidence.

3. Design of the Overall PhC Laser

Most BIC states rely on periodic structures to achieve a high Q factor, which usually requires large PhC structures consisting of 100s of unit cells, and we would like to design structures with fewer than 20 periods to realize miniaturized BIC lasers. First, the Q of BIC states increases with the number of layers, so in region A, structures with enough layers are needed to ensure a high Q factor at 980 nm. Secondly, within region B, the 1550 nm wavelength of light is located in the middle of the bandgap wavelength, and when this structure acts as a reflector for the wavelength of 1550 nm, the more layers there are, the higher the reflectivity is, and the structure with enough layers is needed to ensure a high Q factor at 1550 nm. Finally, between region A to region B, the hole radius size is designed such that the hole radius increases uniformly in the transition region. We have designed structures with different total numbers of layers and different numbers of regions A, B, and transition layers. The total number of layers of Structure 1 is 14, the total number of layers of Structure 2 is 16, the total number of layers of Structures 3 and 4 is 18, and the number of A, B, and transition layers with mode Q factors are shown in Table 1.

Table 1. Q factor of modes with different layer structures.

Structure	Total Layers	Layers of Region A	Layers of Transition Region	Layers of Region B	Q Factor of BIC Mode	Q Factor of M_2 Mode
1	14	6	4	4	8.1×10^1	9.4×10^2
2	16	8	4	4	1.5×10^2	1.8×10^3
3	18	10	4	4	2.3×10^2	3.1×10^3
4	18	12	4	2	3.9×10^2	2.3×10^3

As can be seen from Table 1, the Q factors of both BIC and M_2 modes increase with the increase of the total number of layers. The Q factor of the BIC mode at 980 nm increases significantly with the increase in the number of layers in region A because the Q factor of the BIC mode is directly related to the number of layers of the structure, and the higher the number of layers, the higher the Q factor. Figure 5 shows the variation curve of the Q factor of the BIC mode with the increase in the number of layers in region A.

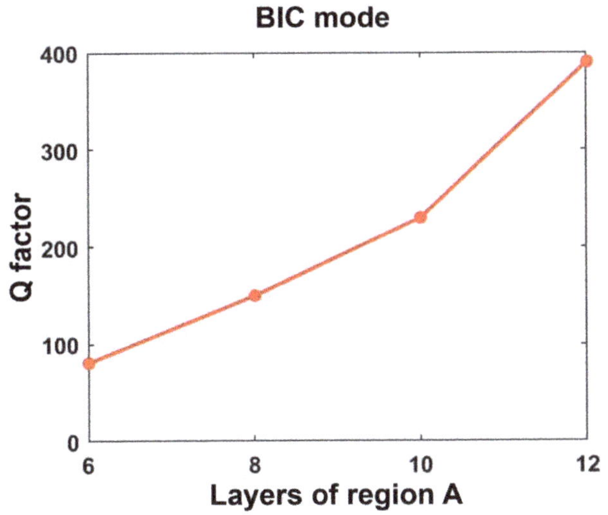

Figure 5. Evolution of the Q of the BIC mode with the number of layers in region A.

When the total number of layers is 18, the larger the number of layers in region B, the larger the reflectivity of the bandgap to 1550 nm light, the better the structure limits the 1550 nm light, and the higher the Q factor of M_2 mode. We simulate the overall structure shown in the table and calculate the eigenfrequencies of the structure near 980 nm and 1550 nm. From Figure 6a–d, it can be seen that the mode field at 980 nm is mainly distributed inside the structure, and the light confinement effect becomes larger with the increase of the number of layers in region A, and the Q factor of the structure also increases. From Figure 6e–h, it can be seen that the mode field at 1550 nm is also confined inside the structure by the mirror structure in region B, and the Q factors increase as the total number of layers and the number of layers in region B increase. Structure 4 has the largest $Q_p \times Q_e = 9.0 \times 10^5$, so it becomes the most desirable structure, at this time, the radii of the innermost 12 layers are equal to 242.3 nm, and the radii of the 13th–16th layers are uniformly increased to 258.7 nm, where r_{13} = 246.4 nm, r_{14} = 250.5 nm, r_{15} = 254.6 nm, r_{16} = 258.7 nm, and the radii of the outermost 17th and 18th layers are equal to 258.7 nm. The Q factors of the BIC state and the M_2 mode are 3.9×10^2 and 2.3×10^3, respectively. A 390-fold reduction in threshold power compared with non-resonant structure at pump wavelength is predicted for this structure. It is important to note that our current design incorporates 18 layers of unit cells, primarily due to computational limitations. Increasing the number of layers beyond this could potentially lead to even higher Q factors for both modes.

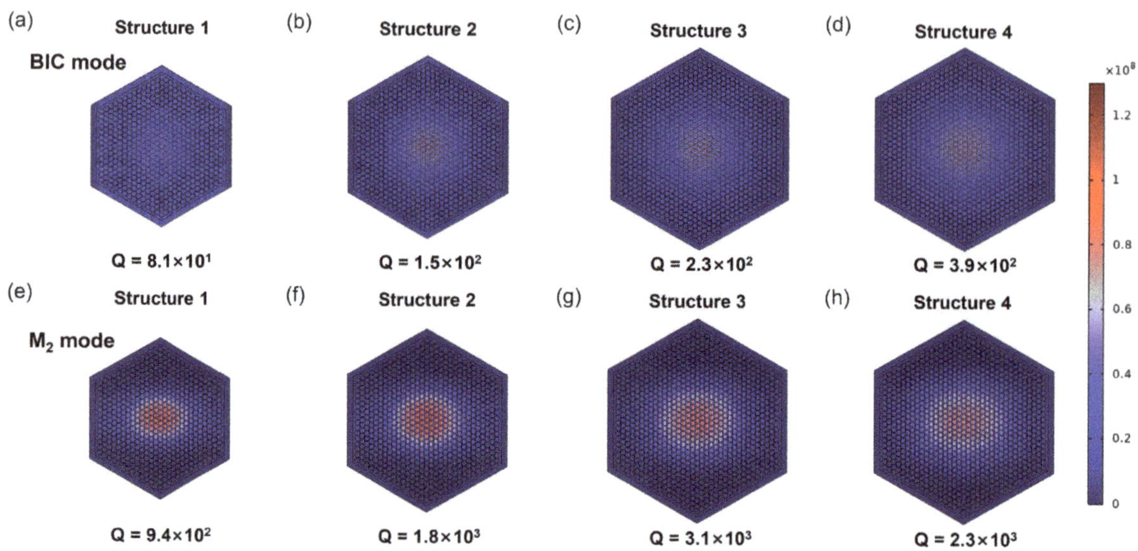

Figure 6. Mode field distribution for PhC structures in Table 1 with different configurations. (**a–d**) BIC modes for the pump light and (**e–h**) M_2 mode for the emitted light in Structure 1–4, respectively.

4. Discussion and Conclusions

The detailed fabrication strategies of our device are as follows: First, a 220-nm layer of $Al_{0.2}Ga_{0.8}As$ is grown via metal-organic chemical vapor deposition (MOCVD) on a 6-inch silicon wafer with a 5 μm SiO_2 buffer layer [38]. Following this, the PhC structure is defined by electron beam lithography (EBL) on an electron-beam resist (ZEP520) and transferred to the $Al_{0.2}Ga_{0.8}As$ film through reactive-ion etching (RIE) in a Cl_2/Ar plasma. Then, 520 nm of Er^{3+}:Al_2O_3 is deposited via plasma-enhanced ALD [34], which simultaneously fills the PhC holes and forms an active upper cladding of around 300 nm thickness. The Er^{3+} concentration can be tuned through the ratio between Er_2O_3 or Al_2O_3 in a supercycle. While carrier diffusion could theoretically reduce active-region carrier density and induce refractive index variations in our Er^{3+}:Al_2O_3-AlGaAs BIC laser design, the actual diffusion into AlGaAs is inherently constrained by two factors: (1) AlGaAs's higher bandgap compared to the pump photon energy minimizes direct excitation-induced carrier generation; (2) localized excitation confines carriers predominantly within the doped regions. So, carrier diffusion should be relatively weak in our design.

In summary, our work presents a novel AlGaAs PhC slab design integrating Er-doped Al_2O_3 as the gain medium. This configuration leverages high-Q modes at both the pump wavelength of 980 nm and the emitting wavelength of 1550 nm, with the former operating as a BIC at the Γ-point and the latter functioning as a PhC cavity mode within the bandgap at the M-point. By achieving calculated Q-factors of 3.9×10^2 and 2.3×10^3 respectively, our design predicts an impressive 390-fold reduction in threshold power compared with non-resonant structure at the pump wavelength. Furthermore, the emitting light's confinement within the PhC cavity mode at the M-point facilitates efficient extraction into in-plane photonic devices, positioning this system as a promising solution for compact on-chip laser applications.

Author Contributions: Conceptualization, L.Z. and X.Z.; validation, Y.M. and J.J.; formal analysis, Y.M.; investigation, Y.M. and X.Z.; resources, L.Z. and X.Z.; data curation, Y.M. and J.J.; writing—original draft preparation, Y.M.; writing—review and editing, L.Z. and X.Z.; supervision, L.Z. and X.Z.; project administration, L.Z. and X.Z. All authors have read and agreed to the published version of the manuscript.

Funding: National Key R&D Program of China (2021YFB2800401).

Institutional Review Board Statement: Not applicable.

Informed Consent Statement: Not applicable.

Data Availability Statement: Data is contained within the article.

Acknowledgments: We acknowledge support by the Advanced Integrated Optoelectronics Facility at Tianjin University.

Conflicts of Interest: The authors declare no conflicts of interest.

References

1. Su, Z.; Li, N.; Salih Magden, E.; Byrd, M.; Purnawirman; Adam, T.N.; Leake, G.; Coolbaugh, D.; Bradley, J.D.; Watts, M.R. Ultra-Compact and Low-Threshold Thulium Microcavity Laser Monolithically Integrated on Silicon. *Opt. Lett.* **2016**, *41*, 5708–5711. [CrossRef] [PubMed]
2. Smit, M.; Williams, K.; Van Der Tol, J. Past, Present, and Future of InP-Based Photonic Integration. *APL Photonics* **2019**, *4*, 050901. [CrossRef]
3. Xiang, C.; Bowers, S.M.; Bjorlin, A.; Blum, R.; Bowers, J.E. Perspective on the Future of Silicon Photonics and Electronics. *Appl. Phys. Lett.* **2021**, *118*, 220501.
4. Zhong, H.; Yu, Y.; Zheng, Z.; Ding, Z.; Zhao, X.; Yang, J.; Wei, Y.; Chen, Y.; Yu, S. Ultra-Low Threshold Continuous-Wave Quantum Dot Mini-BIC Lasers. *Light Sci. Appl.* **2023**, *12*, 100. [CrossRef]
5. Genet, C.; Ebbesen, T.W. Light in Tiny Holes. *Nature* **2007**, *445*, 39–46. [CrossRef]
6. Schuller, J.A.; Barnard, E.S.; Cai, W.; Jun, Y.C.; White, J.S.; Brongersma, M.L. Plasmonics for Extreme Light Concentration and Manipulation. *Nat. Mater.* **2010**, *9*, 193–204. [CrossRef]
7. Ma, R.M.; Oulton, R.F. Applications of Nanolasers. *Nat. Nanotechnol.* **2019**, *14*, 12–22. [CrossRef]
8. Lee, J.; Zhen, B.; Chua, S.L.; Qiu, W.; Joannopoulos, J.D.; Soljačić, M.; Shapira, O. Observation and Differentiation of Unique High-Q Optical Resonances Near Zero Wave Vector in Macroscopic Photonic Crystal Slabs. *Phys. Rev. Lett.* **2012**, *109*, 067401. [CrossRef]
9. Hsu, C.W.; Zhen, B.; Lee, J.; Chua, S.L.; Johnson, S.G.; Joannopoulos, J.D.; Soljačić, M. Observation of Trapped Light within the Radiation Continuum. *Nature* **2013**, *499*, 188–191. [CrossRef]
10. Gao, X.; Yang, L.; Lin, H.; Zhang, L.; Li, J.; Bo, F.; Wang, Z.; Lu, L. Dirac-Vortex Topological Cavities. *Nat. Nanotechnol.* **2020**, *15*, 1012–1018. [CrossRef]
11. Hsu, C.W.; Zhen, B.; Chua, S.L.; Johnson, S.G.; Joannopoulos, J.D.; Soljačić, M. Bloch Surface Eigenstates within the Radiation Continuum. *Light Sci. Appl.* **2013**, *2*, e84. [CrossRef]
12. Hsu, C.W.; Zhen, B.; Stone, A.D.; Joannopoulos, J.D.; Soljačić, M. Bound States in the Continuum. *Nat. Rev. Mater.* **2016**, *1*, 1–13. [CrossRef]
13. Zhen, B.; Hsu, C.W.; Lu, L.; Stone, A.D.; Soljačić, M. Topological Nature of Optical Bound States in the Continuum. *Phys. Rev. Lett.* **2014**, *113*, 257401. [CrossRef] [PubMed]
14. Koshelev, K.; Bogdanov, A.; Kivshar, Y. Meta-Optics and Bound States in the Continuum. *Sci. Bull.* **2019**, *64*, 836–842. [CrossRef]
15. Kang, M.; Mao, L.; Zhang, S.; Xiao, M.; Xu, H.; Chan, C.T. Merging Bound States in the Continuum by Harnessing Higher-Order Topological Charges. *Light Sci. Appl.* **2022**, *11*, 228. [CrossRef] [PubMed]
16. Kodigala, A.; Lepetit, T.; Gu, Q.; Bahari, B.; Fainman, Y.; Kanté, B. Lasing Action from Photonic Bound States in Continuum. *Nature* **2017**, *541*, 196–199. [CrossRef]
17. Zhang, Y.; Chen, A.; Liu, W.; Hsu, C.W.; Wang, B.; Guan, F.; Liu, X.; Shi, L.; Lu, L.; Zi, J. Observation of Polarization Vortices in Momentum Space. *Phys. Rev. Lett.* **2018**, *120*, 186103. [CrossRef]
18. Huang, C.; Zhang, C.; Xiao, S.; Wang, Y.; Fan, Y.; Liu, Y.; Zhang, N.; Qu, G.; Ji, H.; Han, J.; et al. Ultrafast Control of Vortex Microlasers. *Science* **2020**, *367*, 1018–1021. [CrossRef]
19. Yin, X.; Jin, J.; Soljačić, M.; Peng, C.; Zhen, B. Observation of Topologically Enabled Unidirectional Guided Resonances. *Nature* **2020**, *580*, 467–471. [CrossRef]

20. Wang, H.; Zuo, Y.; Yin, X.; Chen, Z.; Zhang, Z.; Wang, F.; Hu, Y.; Zhang, X.; Peng, C. Ultralow-Loss Optical Interconnect Enabled by Topological Unidirectional Guided Resonance. *Sci. Adv.* **2024**, *10*, eadn4372. [CrossRef]
21. Jin, J.; Yin, X.; Ni, L.; Soljačić, M.; Zhen, B.; Peng, C. Topologically Enabled Ultrahigh-Q Guided Resonances Robust to out-of-Plane Scattering. *Nature* **2019**, *574*, 501–504. [CrossRef] [PubMed]
22. Kühne, J.; Wang, J.; Weber, T.; Kühner, L.; Maier, S.A.; Tittl, A. Fabrication Robustness in BIC Metasurfaces. *Nanophotonics* **2021**, *10*, 4305–4312. [CrossRef]
23. Hwang, M.S.; Lee, H.C.; Kim, K.H.; Jeong, K.Y.; Kwon, S.H.; Koshelev, K.; Kivshar, Y.; Park, H.G. Ultralow-Threshold Laser Using Super-Bound States in the Continuum. *Nat. Commun.* **2021**, *12*, 4135. [CrossRef] [PubMed]
24. Ren, Y.; Li, P.; Liu, Z.; Chen, Z.; Chen, Y.L.; Peng, C.; Liu, J. Low-Threshold Nanolasers Based on Miniaturized Bound States in the Continuum. *Sci. Adv.* **2022**, *8*, eade8817. [CrossRef]
25. Baek, S.; Soh, D.B.; Jeong, Y.; Sahu, J.K.; Nilsson, J.; Lee, B. A Cladding-Pumped Fiber Laser with Pump-Reflecting Inner-Cladding Bragg Grating. *IEEE Photonics Technol. Lett.* **2004**, *16*, 407–409. [CrossRef]
26. Raineri, F.; Vecchi, G.; Yacomotti, A.M.; Seassal, C.; Viktorovitch, P.; Raj, R.; Levenson, A. Doubly Resonant Photonic Crystal for Efficient Laser Operation: Pumping and Lasing at Low Group Velocity Photonic Modes. *Appl. Phys. Lett.* **2005**, *86*, 011116. [CrossRef]
27. Xu, C.; Sheng, C.; Zhu, S.; Liu, H. Enhanced Directional Quantum Emission by Tunable Topological Doubly Resonant Cavities. *Opt. Express* **2021**, *29*, 16727–16735. [CrossRef]
28. Li, Y.; Vienne, G.; Jiang, X.; Pan, X.; Liu, X.; Gu, P.; Tong, L. Modeling Rare-earth Doped Microfiber Ring Lasers. *Opt. Express* **2006**, *14*, 7073–7086. [CrossRef]
29. Chremmos, I.; Schwelb, O.; Uzunoglu, N. (Eds.) *Photonic Microresonator Research and Applications*, 1st ed.; Springer: New York, NY, USA, 2010; pp. 3–7.
30. Kim, C.; Ye, C.; Zheng, Y.; Semenova, E.; Yvind, K.; Pu, M. Design and Fabrication of AlGaAs-on-Insulator Microring Resonators for Nonlinear Photonics. *IEEE J. Sel. Top. Quantum Electron.* **2022**, *29*, 9911620. [CrossRef]
31. Mobini, E.; Espinosa, D.H.; Vyas, K.; Dolgaleva, K. AlGaAs Nonlinear Integrated Photonics. *Micromachines* **2022**, *13*, 991. [CrossRef]
32. Agazzi, L.; Bradley, J.D.; Dijkstra, M.; Ay, F.; Roelkens, G.; Baets, R.; Wörhoff, K.; Pollnau, M. Monolithic Integration of Erbium-Doped Amplifiers with Silicon-on-Insulator Waveguides. *Opt. Express* **2010**, *18*, 27703–27711. [CrossRef] [PubMed]
33. Demirtaş, M.; Ay, F. High-Gain Er^{3+}: Al_2O_3 on-Chip Waveguide Amplifiers. *IEEE J. Sel. Top. Quantum Electron.* **2020**, *26*, 3200208. [CrossRef]
34. Zhang, H.; He, Y.; Zhu, S.; Zhou, X.; Zhang, L. Atomic-Layer Engineered Erbium-Doped Waveguide Amplifier with a 14.4 dB Net Gain. *ACS Photonics* **2024**, *12*, 674–683. [CrossRef]
35. Bradley, J.D.B.; Agazzi, L.; Geskus, D.; Ay, F.; Wörhoff, K.; Pollnau, M. Gain Bandwidth of 80 nm and 2 dB/cm Peak Gain in Al_2O_3: Er^{3+} Optical Amplifiers on Silicon. *J. Opt. Soc. Am. B* **2010**, *27*, 187–196. [CrossRef]
36. Zhou, P.; Wang, S.; Wang, X.; He, Y.; Zhou, Z.; Zhou, L.; Wu, K. High-Gain Erbium Silicate Waveguide Amplifier and a Low-Threshold, High-Efficiency Laser. *Opt. Express* **2018**, *26*, 16689–16707. [CrossRef]
37. Wang, X.; Zhou, P.; He, Y.; Zhou, Z. Erbium Silicate Compound Optical Waveguide Amplifier and Laser. *Opt. Mater. Express* **2018**, *8*, 2970–2990. [CrossRef]
38. Papatryfonos, K.; Angelova, T.; Brimont, A.; Reid, B.; Guldin, S.; Smith, P.R.; Tang, M.; Li, K.; Seeds, A.J.; Liu, H.; et al. Refractive Indices of MBE-Grown $Al_xGa_{(1-x)}As$ Ternary Alloys in the Transparent Wavelength Region. *AIP Adv.* **2021**, *11*, 025327. [CrossRef]

photonics

MDPI

Communication

Impact of Non-Vertical Sidewalls on Bandgap Properties of Lithium Niobate Photonic Crystals

Peyman Bagheri [1], Xiaoyan Zhou [1,*] and Lin Zhang [1,2,*]

1 State Key Laboratory of Precision Measuring and Instruments, Key Laboratory of Opto-Electronic Information Technology of Ministry of Education, Tianjin Key Laboratory of Integrated Opto-Electronics Technologies and Devices, School of Precision Instruments and Opto-Electronics Engineering, Tianjin University, Tianjin 300072, China; peymanbagheri@tju.edu.cn

2 Peng Cheng Laboratory, Shenzhen 518038, China

* Correspondence: xiaoyan_zhou@tju.edu.cn (X.Z); lin_zhang@tju.edu.cn (L.Z.)

Abstract: We investigate the influence of non-vertical sidewall angles on the band structure characteristics of thin-film lithium niobate (LN) photonic crystals (PhCs), considering both suspended LN membranes and LN on insulator (LNOI) configurations. Utilizing the gap-to-midgap ratio as a figure-of-merit, we observe a 34% reduction for a suspended LN PhC with 60° sidewall angles compared to the one with vertical sidewalls and a more substantial 73% reduction for LNOI PhCs with 70° sidewall angles. We address this challenge through the optimization of geometrical parameters of PhC unit cells with non-vertical sidewalls, taking fabrication feasibility into account. Our work provides a design guideline for the development of realistic LN PhC devices for future large-scale LN photonic circuits.

Keywords: lithium niobate; photonic crystals; bandgap; slant sidewalls; gap–midgap ratio

1. Introduction

Lithium niobate (LN) stands out as a highly promising material platform for integrated photonics [1,2], which possesses a compelling combination of characteristics, including a broad transparency window, large electro-optic and nonlinear effects, and low absorption [3]. In recent years, the substantial development in thin-film LN photonics [4–6] has been noteworthy. This has led to the demonstration of low passive waveguide loss down to dB/m [7,8] and fast active electro-optical (EO) modulators up to 100 GHz [9–11], underscoring the great potential of LN photonics for large-scale and high-performance photonic integrated circuits (PICs).

Despite significant progress in LN photonics, a notable challenge in constructing scalable LN circuits lies in the substantial device size, which partially stems from the preference for ridge waveguides in many cases due to the need for electrodes in EO modulators [9–12], the difficulty in etching LN films [13], and the requirements for low-loss waveguides [7,8]. In this context, photonic crystal (PhC) structures, guiding optical modes through the photonic bandgap rather than total internal reflection, offer strong light confinement and would thus be beneficial for shrinking the devices. LN PhC structures on thin-film platforms are preferred over conventional thin-film lithium niobate photonic structures primarily due to their exceptional capability to control and tightly confine light at the sub-wavelength scale [14,15]. This strong confinement leads to a significantly enhanced electro-optic interaction and tuning efficiency, enabling devices with a tiny electro-optic modal volume and consequently high energy efficiency. Furthermore, this tight confinement can result in a broad modulation bandwidth that can reach the photon-lifetime limit of the cavity, making photonic crystal structures highly advantageous for miniaturized, high-speed, and energy-efficient integrated photonic circuits [16]. Representative works on LN platforms include PhC cavities on a wavelength scale [16,17], sharp waveguides bends [18], and modulators with enhanced EO interactions [14,19,20].

Most LN fabrication techniques yield a sidewall angle (θ) less than 90°, including diamond dicing [21,22], chemical mechanical polishing [8], a focused ion beam [23–26], and dry etching [7,9–12,14,16–19].

The sidewall angle typically falls within the range of 40° to 80° [1,11,16–18,27], depending on the specific fabrication methods. This often results in cone-like holes in fabricated LN PhCs, as illustrated in Figure 1a,b. To address this, the limitation of slanted sidewalls has been considered in the design of the PhCs [17]. Limited aspect ratios due to the conical shapes of holes are typically observed when PhC structures are produced in annealed proton exchanged (APE) waveguides by focused ion beam (FIB) milling [28]. In contrast, advanced dry etching techniques such as inductively coupled plasma (ICP) etching and atomic layer etching (ALE) have been explored to achieve steeper sidewalls and smoother surfaces [29], requiring precise control over plasma parameters and etching conditions. The slanted sidewalls of these holes [16,23,25,27] may lead to performance degradation of the PhCs. For PhC waveguides built from slanted holes, they may also lead to mode hybridization and mode conversion. Also, the cone-like holes may impede the fabrication of PhCs with a small lattice constant due to structural interference. However, there have been few reports on the impact of non-vertical sidewalls on LN PhC devices, and the potential mitigation of this issue with practical sidewall angles remains unexplored. Moreover, the LN PhCs with slanted sidewalls offer a novel avenue for studying bound states in the continuum (BICs), which are peculiar states enabling light trapping in thin photonic structures [30–32]. The slanted geometry of the etched sidewalls facilitates the creation of intrinsic BICs, which can be excited under vertical incidence conditions [33]. Thus, the role of sidewall angles in LN PhC devices is believed to be important in many practical application scenarios.

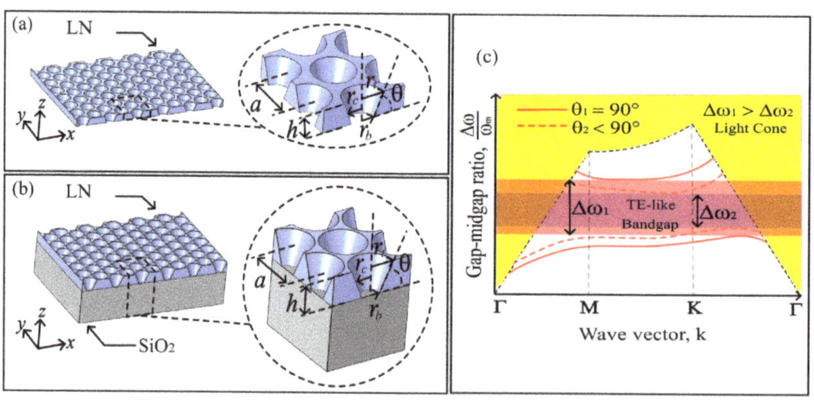

Figure 1. LN PhC slabs consisting of air holes with slanted sidewalls and the PhC band structure. (**a**,**b**) Schematic of hexagonal (**a**) LN PhC slab suspended in the air and (**b**) LNOI PhC slab with cone-like holes. (**c**) Band structures for the TE-like mode in LN PhC slabs, showing a reduced bandgap ($\Delta\omega_2 < \Delta\omega_1$) for slanted sidewall angle ($\theta_2 < \theta_1 = 90°$).

In this paper, we study the impact of sidewall angle on the bandgap characteristics for thin-film LN PhCs with a hexagonal lattice for suspended LN slabs and LN on insulator (LNOI) slabs, as shown in Figure 1a,b. It is found that the sidewall angle critically determines the bandgap of LN PhCs. In some cases, particularly for LNOI PhC devices, the bandgap, $\Delta\omega$, may even be closed. The gap-to-midgap ratio, $\frac{\Delta\omega}{\omega_m}$, is utilized as a figure-of-merit [34] to evaluate the bandgap properties in PhCs with slant sidewalls. The central wavelength of the bandgap, a size-dependent characteristic, can be tuned by adjusting the main geometrical parameters of PhCs. This value exhibited a range of variation from 1338 nm to 1717 nm under different parameter modification conditions and for configurations explored in our work. For suspended LN PhCs, reducing the sidewall angle to θ = 60° results in a 34% reduction in the gap–midgap ratio compared to that with vertical sidewalls. Similarly, for LN PhCs on silica, decreasing θ to 70° leads to a 73% reduction in $\frac{\Delta\omega}{\omega_m}$. To address this decrease, we propose compensative solutions, such as increasing the central hole radius, r_c, or reducing the lattice constant, a, and the slab thickness, h. Especially, we note that a reduction in h by 150 nm increases $\frac{\Delta\omega}{\omega_m}$ by 34% for a suspended LN PhC with θ = 60° and h = 400 nm. Similar optimization in h for an LNOI PhC slab with θ = 70° yields a ∼200% increment. Our study establishes design guidelines for PhC devices on LN, bridging practical fabrication needs with optical performance demands. This facilitates energy-efficient, compact LN devices robust to process variations, enabling scalable photonic integrated circuits where optimized architectures align with achievable fabrication precision. The guidelines advance high-performance LN photonics by embedding manufacturability constraints directly into device design paradigms.

2. Design and Characterization

Utilizing hexagonal array of holes in LN slabs, we studied TE-like bandgap properties by changing sidewall angles. This structure was selected as the representative of all other possible structures as the bandgap size in the considered structure is a large value, which enables us to show bandgap properties controlled by sidewall angles in detail. On the other hand, considering the popularity of compact LN photonic devices based on TE mode structures, the TE-like band structure was selected as the reference for evaluation. First, we investigated semi-2D LN PhC slabs. Due to the absence of the bandgap for LN PhC slabs thinner than 1 µm with a square lattice, we focused on the hexagonal lattice. The unit cell comprises an air hole within the LN thin film for the both suspended LN in air and LNOI configurations, as shown in Figure 1a and 1b, respectively. To determine the band structure within the Brillouin zone for unit cells in these LN PhCs, a 3D eigenmode solver based on the finite-element method was employed. The model was surrounded with symmetric boundaries in periodic directions and the perfectly matched layer (PML) at cladding area boundaries. Fine mesh size (as small as one-eighth of effective wavelength) was applied to the model to obtain proper accuracy in the extracted eigenmode resolution. Quasi-TE/TM modes were distinguished via analyzing electric field components. A custom algorithm was utilized for detecting TE-like bandgaps and calculating the gap-to-midgap ratio from exported eigenvalues and their corresponding dominant field component. Finally, the eigenvalues were plotted over the normalized wave vector. a, h, and r_c are adjusted so that the bandgap is around the telecom bands. Here, the slanted sidewalls impose geometrical limitations on the design of PhCs. When $\theta < 90°$, the hole radius at the top of a unit cell, r_t, is larger than r_c, potentially causing overlap between air holes if $r_t > \frac{\sqrt{3}}{2}a$. Consequently, the geometrical parameters of unit cells (a, h, and r_c) must be carefully adjusted to avoid structural interference caused by slanted sidewalls.

The band structures for suspended LN PhC and LNOI PhC are obtained for different sidewall angles in Figure 2a and 2b, respectively. Both TE-like and TM-like bands are presented. Distinct sets of parameters a, h, and r_c are employed for LN PhC and LNOI PhC to align the bandgap around the telecom bands. By setting a = 620 nm, r_c = 180 nm, and h = 300 nm in the suspended LN PhC slab, the bandgap between the two lowest TE-like bands (labeled as Band 1 and Band 2 in Figure 2a is in the range of 1350 nm to 1530 nm.

The bandgap is between 1540 nm and 1630 nm for a = 580 nm, r_c = 140 nm, and h = 450 nm for the case of the LNOI PhC slab (Figure 2b). These parameters remain constant for varied θ, and we note that for $\theta = 60°$, the LNOI PhC has structural interference. These two sets of parameters are used throughout the paper unless specified otherwise.

In Figure 2, the suspended LN PhC features a large TE-like bandgap (referred to as $\Delta\omega$) that is desirable for building PhC waveguides or cavities. In contrast, $\Delta\omega$ is greatly reduced in LNOI PhC slabs due to a reduced refractive index contrast between LN and the substrate. A higher-index substrate also leads to a downward shift in the light cone, bounded by $\omega = \frac{kc}{n_{silica}}$, where n_{silica} is the refractive index of silica, thereby imposing constraints in the potentially achievable bandgap. In both cases, $\Delta\omega$ is reduced when θ decreases, as shown by the yellow dashed lines in Figure 2a,b.

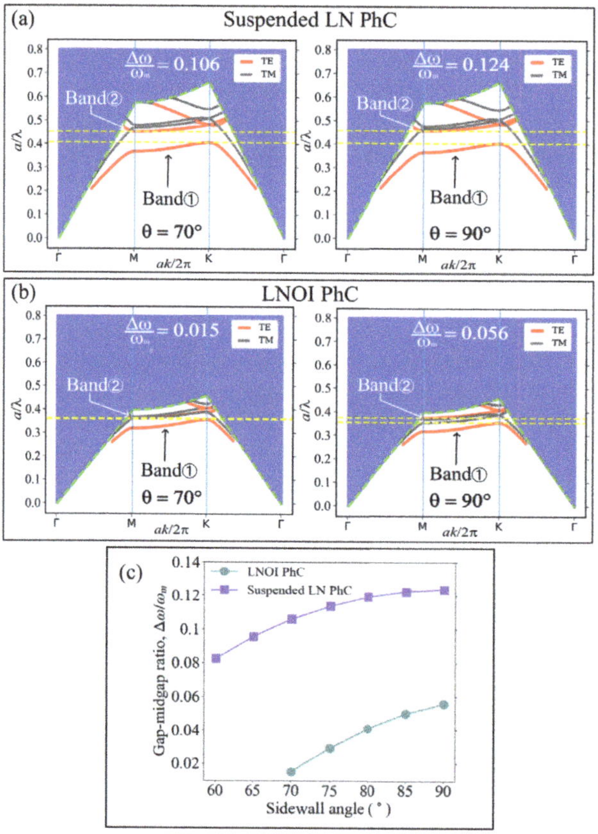

Figure 2. Band structures and $\frac{\Delta\omega}{\omega_m}$ with varied θ for (**a**) suspended LN PhC slabs, with a = 620 nm, r_c = 180 nm, and h = 300 nm, and for (**b**) LNOI PhC slabs with a = 580 nm, r_c = 140 nm, and h = 450 nm. In both (**a**,**b**), the blue area denotes the light cone and yellow dashed lines show the bandgap. (**c**) $\frac{\Delta\omega}{\omega_m}$ increases with θ.

The gap–midgap ratio, $\frac{\Delta\omega}{\omega_m}$, offers an independent metric across geometric scales to quantify the performance of PhCs with different sidewall angles, where ω_m is the angular frequency at the midpoint of the TE-like bandgap. A small $\frac{\Delta\omega}{\omega_m}$ is generally undesirable and indicates the degraded "light-confining capability" of PhCs. $\frac{\Delta\omega}{\omega_m}$ for varied θ values is annotated in Figure 2a,b and summarized in Figure 2c, revealing a notable reduction when $\theta < 90°$. Specifically, $\frac{\Delta\omega}{\omega_m}$ drops from 0.124 at $\theta = 90°$ to 0.082 at $\theta = 60°$ for the suspended PhC slab, marking a reduction of ~34%. For the LNOI PhC, and $\frac{\Delta\omega}{\omega_m}$ decreases more severely by 73% from 0.056 at $\theta = 90°$ to 0.015 at $\theta = 70°$. In addition, $\frac{\Delta\omega}{\omega_m}$ is more sensitive to sidewall angle variation when θ is small, as in Figure 2c for both PhC configurations. In the suspended case, $\frac{\Delta\omega}{\omega_m}$ decreases by around $\sim$14% as θ decreases by 5° from $\theta = 65°$, whereas $\frac{\Delta\omega}{\omega_m}$ is almost unchanged for $\theta > 85°$. This variation in $\frac{\Delta\omega}{\omega_m}$ is more significant for LNOI PhCs. $\frac{\Delta\omega}{\omega_m}$ decreases by >45% as θ decreases by 5° from $\theta = 75°$, while the $\frac{\Delta\omega}{\omega_m}$ is reduced by $\sim$10% with θ from 90° to 85°. $\Delta\omega$ becomes 0 for sidewall angles smaller than 65°.

The decrease in $\frac{\Delta\omega}{\omega_m}$ for non-vertical sidewalls is jointly caused by a slight decrease in the eigenfrequencies in Band 1, the dielectric band, and a large decrease in the eigenfrequencies in Band 2, the air band. This can be understood from the mode distributions for Band 1 at the K-point and Band 2 at the M-point in the unit cell for various sidewall angles, as shown in Figure 3. For both PhC configurations, when θ decreases from 90° to 70°, it is noted that the normalized electric field distribution is intensified around the bottom inside the air holes for Band 2, in comparison to a slight change for the electric field inside the dielectric area for Band 1. Therefore, one expects to observe a more pronounced variation in eigenfrequency for Band 2 compared to Band 1 as θ decreases. This unequal eigenfrequency shift of the two bands results in a reduced bandgap and a significant decrease in $\frac{\Delta\omega}{\omega_m}$ for a small sidewall angle.

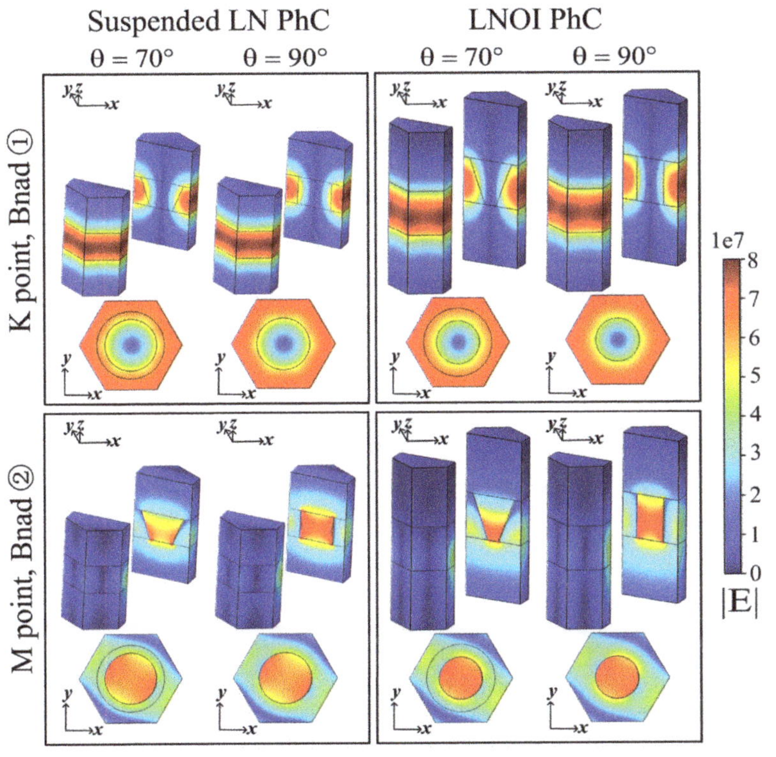

Figure 3. Electric field distribution around LN PhC slab region in the unit cell at K point on Band 1 and M point on Band 2 for $\theta = 70°$ and $\theta = 90°$.

3. Manipulating the Impact of Sidewall Angles

We examine the effect of the sidewall angle on $\frac{\Delta\omega}{\omega_m}$ in both types of LN PhCs for varied a, r_c, or h and optimize them to maximize $\frac{\Delta\omega}{\omega_m}$ for practical sidewall angles. Figure 4 shows that the overall $\frac{\Delta\omega}{\omega_m}$ increases with r_c for varied θ. For suspended LN PhC in Figure 4a, the changes in $\frac{\Delta\omega}{\omega_m}$ with r_c are shown for $60° \leq \theta \leq 90°$. As r_c increases from 150 nm to 210 nm,

$\frac{\Delta\omega}{\omega_m}$ increases from 0.069 to 0.167 for $\theta = 90°$, marking a 142% rise. Similarly, for $\theta = 60°$, the ratio increases from 0.041 to 0.113, reflecting a larger increment of 175%. These results indicate that $\frac{\Delta\omega}{\omega_m}$ features a more pronounced increase with r_c at a smaller θ. Figure 4b shows the $\frac{\Delta\omega}{\omega_m}$ for LNOI PhCs with $65° \leq \theta \leq 90°$ by changing r_c. Compared to Figure 4a, adding a silica sublayer to the PhC slab causes a remarkable decrement in $\frac{\Delta\omega}{\omega_m}$. Despite this change, the overall variations in $\frac{\Delta\omega}{\omega_m}$ with r_c for LNOI PhCs are similar to those observed in suspending LN PhC slabs. $\frac{\Delta\omega}{\omega_m}$ increases from 0.019 to 0.085 at $\theta = 90°$ when r_c increases from 120 nm to 160 nm. Similarly, in the same range of r_c, the $\frac{\Delta\omega}{\omega_m}$ ratio shows an increment from 0.0003 to 0.054 at $\theta = 75°$, showing a greater impact of r_c on $\frac{\Delta\omega}{\omega_m}$ at smaller θ, which is also similar to the observed behavior in suspended LN PhCs.

The increase in $\frac{\Delta\omega}{\omega_m}$ with r_c can be leveraged to counteract the impact of non-vertical sidewall on $\frac{\Delta\omega}{\omega_m}$. For example, $\frac{\Delta\omega}{\omega_m}$ for LNOI PhC drops from 0.038 at $\frac{\Delta\omega}{\omega_m} = 90°$ to 0.003 at $\frac{\Delta\omega}{\omega_m} = 70°$ when $r_c = 130$ nm in Figure 4b, which can be improved to 0.037 with a larger $r_c = 160$ nm. In this sense, it is desirable to design the unit cell with a large r_c or, more precisely, r_c/a, for both suspended LN and LNOI PhC slabs. In the case of LNOI PhCs, it should be noted that r_c cannot be excessively large. This limitation arises because an overly large r_c would lead to Band 2 being positioned inside the light cone. In addition, the possible r_c value for a fixed lattice constant is limited in fabrication. This limitation becomes tighter for PhCs with slanted sidewalls, which limits the achievable $\frac{\Delta\omega}{\omega_m}$.

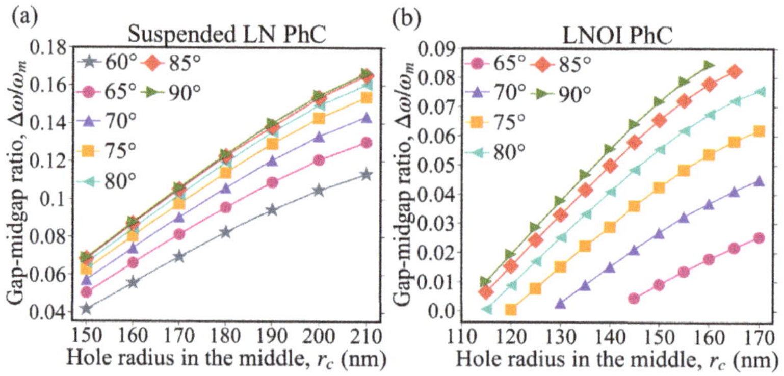

Figure 4. The impact of changing θ on $\frac{\Delta\omega}{\omega_m}$ for varied r_c in (**a**) suspended LN PhCs with $a = 620$ nm and $h = 300$ and (**b**) LNOI PhCs with $a = 580$ nm and $h = 450$ nm. Omitted points are either due to the absence of a bandgap or the placement of Band 2 within the light cone.

An increase in a, with fixed r_c and h values, results in a reduction in $\frac{\Delta\omega}{\omega_m}$ for both suspended LN PhCs and LNOI PhCs, as shown in Figure 5. Specifically, Figure 5a shows that $\frac{\Delta\omega}{\omega_m}$ decreases when a increases from 600 nm to 640 nm for $60° \leq \theta \leq 90°$ in suspended PhC slabs. The change in $\frac{\Delta\omega}{\omega_m}$ with a is slightly different at various θ. For $\theta = 90°$, $\frac{\Delta\omega}{\omega_m}$ falls by >17%, reducing from 0.136 for $a = 600$ nm to 0.112 for $a = 640$ nm, which is 5% more than the variation in $\frac{\Delta\omega}{\omega_m}$ for $\theta = 60°$, reducing from 0.089 for $a = 600$ nm to 0.076 for $a = 640$ nm. On the other hand, Figure 5b shows $\frac{\Delta\omega}{\omega_m}$ in the LNOI configuration as a function of a in the range of 540 nm to 620 nm for different θ. $\frac{\Delta\omega}{\omega_m}$ for $a = 620$ nm is half of that for $a = 540$ nm in the case of $\theta = 90°$. For $\theta = 70°$, $\frac{\Delta\omega}{\omega_m}$ is reduced by more than 75% comparing $a = 540$ nm to $a = 620$ nm. The bandgap is almost closed for $\theta = 65°$, leading to $\frac{\Delta\omega}{\omega_m} \approx 0$. Similar to suspended PhCs, $\frac{\Delta\omega}{\omega_m}$ is more sensitive to a at a smaller θ.

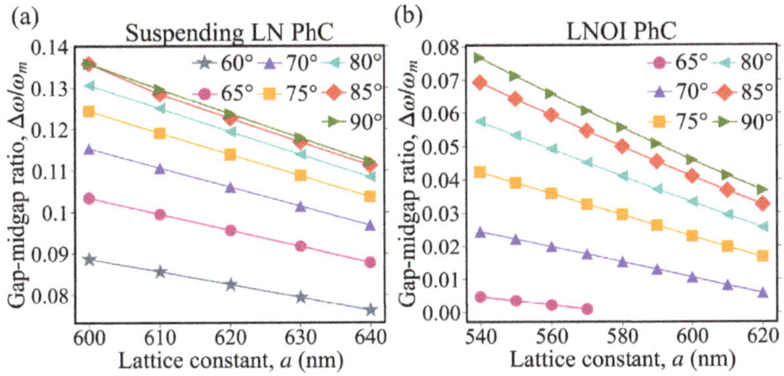

Figure 5. The impact of changing θ on $\frac{\Delta\omega}{\omega_m}$ by varied a for (**a**) suspended PhCs with r_c = 180 nm and h = 300 nm and (**b**) LNOI PhCs with r_c = 140 nm and h = 450 nm. Omitted points are due to the absence of a bandgap.

Accordingly, by reducing a, the PhC with a small θ could have a comparable $\frac{\Delta\omega}{\omega_m}$ for a PhC with a vertical sidewall angle and a large a. For example, comparing $\frac{\Delta\omega}{\omega_m}$ values in Figure 5a, a suspended PhC slab in the air with θ = 80° and a = 630 nm can be replaced by one with θ = 70° and a = 600 nm to obtain the same $\frac{\Delta\omega}{\omega_m}$ if the working wavelength is flexible. Similarly, for the LNOI PhC in Figure 5b, a can be reduced by 60 nm from a = 600 nm to compensate for the 15° reduction in θ from 90° to 75° to obtain the same $\frac{\Delta\omega}{\omega_m}$.

By modifying h while maintaining a constant r_c and a, both the dielectric area and the air holes are affected. This leads to a more intricate variation in $\frac{\Delta\omega}{\omega_m}$ with h, as shown in Figure 6a,b for the suspended LN PhC slab and the LNOI PhC slab, respectively. The most notable feature observed in $\frac{\Delta\omega}{\omega_m}$ concerning h is the distinct variations it exhibits for θ values within different ranges. For suspended LN PhC slabs shown in Figure 6a, $\frac{\Delta\omega}{\omega_m}$ increases monotonically with h for $80° \leq \theta \leq 90°$, showing a slower rate of increase towards h = 400 nm. For a smaller θ in the range of $60° \leq \theta \leq 75°$, it is very interesting that $\frac{\Delta\omega}{\omega_m}$ of the PhC features an optimized value at a specific h. For θ = 65°, it appears evident that the commonly used 300 nm thick LN slab yields the largest $\frac{\Delta\omega}{\omega_m}$, thus showcasing a superior light-confining capability for constructing functional devices. In addition, Figure 6a highlights the increased sensitivity of $\frac{\Delta\omega}{\omega_m}$ to sidewall angle when thicker slabs are adopted.

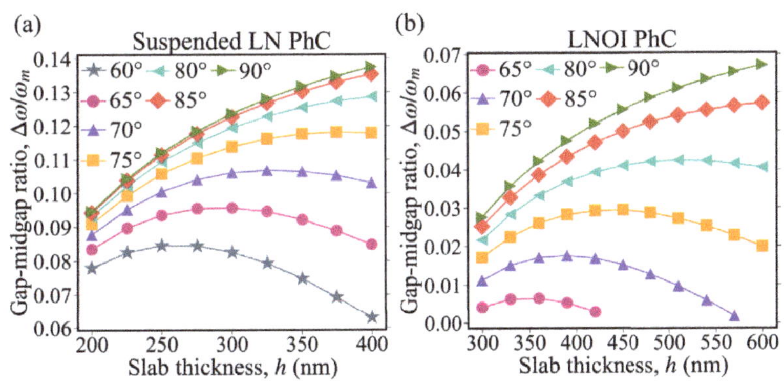

Figure 6. The impact of changing θ on $\frac{\Delta\omega}{\omega_m}$ by varied h for (**a**) suspended LN PhCs with r_c = 180 nm and a = 620 nm and (**b**) LNOI PhCs with r_c = 140 nm and a = 580 nm. Omitted points are due to the absence of a bandgap.

For h = 200 nm, $\frac{\Delta\omega}{\omega_m}$ decreases from 0.095 to 0.078, marking an approximately 18% decrement, while this decrement rises to ~54% for h = 400 nm. Similarly, $\frac{\Delta\omega}{\omega_m}$ for LNOI PhCs in Figure 6b increases monotonically with h between 300 nm and 600 nm for $85^\circ \leq \theta \leq 90^\circ$ and shows a maximum value at a specified h for a smaller θ, i.e., $65^\circ \leq \theta \leq 80^\circ$. In the case of $\theta = 65^\circ$, the bandgap tends to diminish, existing only for 300 nm $\leq h \leq$ 420 nm, with a maximum value of 0.0065 achieved around h = 360 nm. Figure 6a,b could serve as a reference for optimizing h with the constraints of fabrication-limited θ. This is particularly crucial for LNOI PhCs due to the substantial variations in $\frac{\Delta\omega}{\omega_m}$ concerning θ.

The optimal geometrical parameters for achieving photonic bandgaps within the communication band (around λ = 1550 nm) are not unique and can be systematically selected based on fabrication limitations or application-specific design criteria. For suspended LN PhC unit cell, a bandgap encompassing the communication band is achieved with the following parameters: h = 400 nm, r_c = 180 nm, a = 620 nm, and sidewall angle $\theta = 60^\circ$ which can be proposed as one of optimized configurations. In contrast, the LNOI PhC unit cell exhibits comparable performance with distinct optimized geometrical characteristics, with h = 450 nm, r_c = 165 nm, a = 580 nm, and $\theta = 70^\circ$. Both configurations maintain sidewall angles within experimentally feasible ranges, underscoring their practical viability for scalable device integration.

4. Conclusions

In this work, we investigated the impact of non-vertical sidewalls on the bandgap characteristics of thin-film LN PhCs. The gap–midgap ratio is used as a metric for comparing the performance of PhCs with different sidewall angles. This study underscores the inherent drawback associated with the formation of slanted sidewalls in LN PhC devices, especially for LNOI PhCs. Furthermore, we conducted a comprehensive geometrical optimization of PhC unit cells to counteract the reduction in the gap–midgap ratio induced by angled sidewalls and provide a guideline for the fabrication-aware design of LN PhC devices, extending the scalability of LN photonics.

Author Contributions: Conceptualization, P.B. and L.Z.;methodology, P.B. and X.Z.; software, P.B.; formal analysis, P.B., X.Z. and L.Z.; data curation, P.B. and X.Z.; writing—original draft preparation, P.B. and X.Z.; supervision, writing—review and editing, L.Z. All authors have read and agreed to the published version of the manuscript.

Funding: This research was funded by Innovation Program for Quantum Science and Technology (2024ZD0300800).

Institutional Review Board Statement: Not applicable.

Informed Consent Statement: Not applicable.

Data Availability Statement: The data that support the findings of this study are available from the corresponding author upon reasonable request.

Conflicts of Interest: The authors declare no conflicts of interest.

References

1. Zhu, D.; Shao, L.; Yu, M.; Cheng, R.; Desiatov, B.; Xin, C.; Hu, Y.; Holzgrafe, J.; Ghosh, S.; Shams-Ansari, A.; et al. Integrated photonics on thin-film lithium niobate. *Adv. Opt. Photonics* **2021**, *13*, 242–352. [CrossRef]
2. Qi, Y.; Li, Y. Integrated lithium niobate photonics. *Nanophotonics* **2020**, *9*, 1287–1320. [CrossRef]
3. Weis, R.; Gaylord, T. Lithium niobate: Summary of physical properties and crystal structure. *Appl. Phys. A* **1985**, *37*, 191–203. [CrossRef]
4. Jia, Y.; Wang, L.; Chen, F. Ion-cut lithium niobate on insulator technology: Recent advances and perspectives. *Appl. Phys. Rev.* **2021**, *8*, 011307. [CrossRef]

5. Poberaj, G.; Hu, H.; Sohler, W.; Guenter, P. Lithium niobate on insulator (LNOI) for micro-photonic devices. *Laser Photonics Rev.* **2012**, *6*, 488–503. [CrossRef]
6. Honardoost, A.; Abdelsalam, K.; Fathpour, S. Rejuvenating a versatile photonic material: Thin-film lithium niobate. *Laser Photonics Rev.* **2020**, *14*, 2000088. [CrossRef]
7. Zhang, M.; Wang, C.; Cheng, R.; Shams-Ansari, A.; Lončar, M. Monolithic ultra-high-Q lithium niobate microring resonator. *Optica* **2017**, *4*, 1536–1537. [CrossRef]
8. Wu, R.; Wang, M.; Xu, J.; Qi, J.; Chu, W.; Fang, Z.; Zhang, J.; Zhou, J.; Qiao, L.; Chai, Z.; et al. Long low-loss-litium niobate on insulator waveguides with sub-nanometer surface roughness. *Nanomaterials* **2018**, *8*, 910. [CrossRef]
9. Arab Juneghani, F.; Gholipour Vazimali, M.; Zhao, J.; Chen, X.; Le, S.T.; Chen, H.; Ordouie, E.; Fontaine, N.K.; Fathpour, S. Thin-Film Lithium Niobate Optical Modulators with an Extrapolated Bandwidth of 170 GHz. *Adv. Photonics Res.* **2023**, *4*, 2200216. [CrossRef]
10. Wang, C.; Zhang, M.; Chen, X.; Bertrand, M.; Shams-Ansari, A.; Chandrasekhar, S.; Winzer, P.; Lončar, M. Integrated lithium niobate electro-optic modulators operating at CMOS-compatible voltages. *Nature* **2018**, *562*, 101–104. [CrossRef]
11. He, M.; Xu, M.; Ren, Y.; Jian, J.; Ruan, Z.; Xu, Y.; Gao, S.; Sun, S.; Wen, X.; Zhou, L.; et al. High-performance hybrid silicon and lithium niobate Mach–Zehnder modulators for 100 Gbit s- 1 and beyond. *Nat. Photonics* **2019**, *13*, 359–364. [CrossRef]
12. Zhang, Y.; Shen, J.; Li, J.; Wang, H.; Feng, C.; Zhang, L.; Sun, L.; Xu, J.; Liu, M.; Wang, Y.; et al. High-speed electro-optic modulation in topological interface states of a one-dimensional lattice. *Light. Sci. Appl.* **2023**, *12*, 206. [CrossRef]
13. Mookherjea, S.; Mere, V.; Valdez, F. Thin-film lithium niobate electro-optic modulators: To etch or not to etch. *Appl. Phys. Lett.* **2023**, *122*, 120501. [CrossRef]
14. Li, M.; Ling, J.; He, Y.; Javid, U.A.; Xue, S.; Lin, Q. Lithium niobate photonic-crystal electro-optic modulator. *Nat. Commun.* **2020**, *11*, 4123. [CrossRef]
15. Wang, Y.; Yao, Y.; Zhang, H.; Liu, B.; Duan, S.; Lin, W. An electrically controlled tunable photonic crystal filter based on thin-film lithium niobate. *Optoelectron. Lett.* **2024**, *20*, 200–204. [CrossRef]
16. Ge, R.; Yan, X.; Liang, Z.; Li, H.; Wu, J.; Liu, X.; Chen, Y.; Chen, X. Large quality factor enhancement based on cascaded uniform lithium niobate bichromatic photonic crystal cavities. *Opt. Lett.* **2023**, *48*, 113–116. [CrossRef]
17. Liang, H.; Luo, R.; He, Y.; Jiang, H.; Lin, Q. High-quality lithium niobate photonic crystal nanocavities. *Optica* **2017**, *4*, 1251–1258. [CrossRef]
18. Shi, C.; Yuan, J.; Luo, X.; Shi, S.; Lu, S.; Yuan, P.; Xu, W.; Chen, Z.; Yu, H. Transmission characteristics of multi-structure bandgap for lithium niobate integrated photonic crystal and waveguide. *Opt. Commun.* **2020**, *461*, 125222. [CrossRef]
19. Chen, G.; Wang, H.; Chen, B.; Ruan, Z.; Guo, C.; Chen, K.; Liu, L. Compact slow-light waveguide and modulator on thin-film lithium niobate platform. *Nanophotonics* **2023**, *12*, 3603–3611. [CrossRef]
20. Zhang, J.; Pan, B.; Liu, W.; Dai, D.; Shi, Y. Ultra-compact electro-optic modulator based on etchless lithium niobate photonic crystal nanobeam cavity. *Opt. Express* **2022**, *30*, 20839–20846. [CrossRef]
21. Wang, S.; Zhao, J.; Gu, J.; Bu, M.; Fan, L.; Li, S.; Qin, X.; Yao, Y.; Ren, Y.; Wang, L. $LiNbO_3$ channel and ridge waveguides based on helium ion implantation combined with lithography and precise diamond dicing. *Chin. Opt. Lett.* **2022**, *20*, 071301. [CrossRef]
22. Volk, M.F.; Suntsov, S.; Rüter, C.E.; Kip, D. Low loss ridge waveguides in lithium niobate thin films by optical grade diamond blade dicing. *Opt. Express* **2016**, *24*, 1386–1391. [CrossRef] [PubMed]
23. Si, G.; Danner, A.J.; Teo, S.L.; Teo, E.J.; Teng, J.; Bettiol, A.A. Photonic crystal structures with ultrahigh aspect ratio in lithium niobate fabricated by focused ion beam milling. *J. Vac. Sci. Technol. B* **2011**, *29*, 021205. [CrossRef]
24. Burr, G.; Diziain, S.; Bernal, M.P. The impact of finite-depth cylindrical and conical holes in lithium niobate photonic crystals. *Opt. Express* **2008**, *16*, 6302–6316. [CrossRef]
25. Cai, L.; Han, H.; Zhang, S.; Hu, H.; Wang, K. Photonic crystal slab fabricated on the platform of lithium niobate-on-insulator. *Opt. Lett.* **2014**, *39*, 2094–2096. [CrossRef]
26. Lin, J.; Yao, N.; Hao, Z.; Zhang, J.; Mao, W.; Wang, M.; Chu, W.; Wu, R.; Fang, Z.; Qiao, L.; et al. Broadband quasi-phase-matched harmonic generation in an on-chip monocrystalline lithium niobate microdisk resonator. *Phys. Rev. Lett.* **2019**, *122*, 173903. [CrossRef]
27. Chen, D.; Muhammad, S.; Huang, W.; Zheng, X.; Wen, G.; Huang, Y. Parameter investigations on lithium-niobate-based photonic crystal optomechanical cavity. *Results Phys.* **2023**, *48*, 106458. [CrossRef]
28. Geiss, R.; Diziain, S.; Iliew, R.; Etrich, C.; Hartung, H.; Janunts, N.; Schrempel, F.; Lederer, F.; Pertsch, T.; Kley, E.B. Light propagation in a free-standing lithium niobate photonic crystal waveguide. *Appl. Phys. Lett.* **2010**, *97*, 131109. [CrossRef]
29. Shen, B.; Hu, D.; Dai, C.; Yu, X.; Tan, X.; Sun, J.; Jiang, J.; Jiang, A. Advanced etching techniques of $LiNbO_3$ nanodevices. *Nanomaterials* **2023**, *13*, 2789. [CrossRef]
30. Zhang, X.; Liu, Y.; Han, J.; Kivshar, Y.; Song, Q. Chiral emission from resonant metasurfaces. *Science* **2022**, *377*, 1215–1218. [CrossRef]

31. Wang, J.; Li, P.; Zhao, X.; Qian, Z.; Wang, X.; Wang, F.; Zhou, X.; Han, D.; Peng, C.; Shi, L.; et al. Optical bound states in the continuum in periodic structures: Mechanisms, effects, and applications. *Photonics Insights* **2024**, *3*, R01. [CrossRef]
32. He, W.; Wang, Y. Enhancement of second-harmonic generation in a lithium niobate metasurface by exploring the bound states in the continuum. *Opt. Express* **2024**, *32*, 39415–39428. [CrossRef] [PubMed]
33. Chen, Y.; Deng, H.; Sha, X.; Chen, W.; Wang, R.; Chen, Y.H.; Wu, D.; Chu, J.; Kivshar, Y.S.; Xiao, S.; et al. Observation of intrinsic chiral bound states in the continuum. *Nature* **2023**, *613*, 474–478. [CrossRef] [PubMed]
34. Joannopoulos, J.D.; Johnson, S.G.; Winn, J.N.; Meade, R.D. Photonic Crystals. In *Molding the Flow of Light*, 2nd ed.; Princeton University Press: Princeton, NJ, USA, 2008. [CrossRef]

photonics

MDPI

Communication

Longitudinal Polarization Vortices Generated via Terahertz Ring Resonator

Mingyu Ji †, Tengjiao Wang † and Jingya Xie *

Shanghai Key Lab of Modern Optical System, University of Shanghai for Science and Technology, No. 516 JunGong Road, Shanghai 200093, China
* Correspondence: xiejy@usst.edu.cn
† These authors contributed equally to this work.

Abstract: Vortex beams characterized by helical phase wavefronts enable innovative explorations of optical and physical interactions. This work experimentally realizes longitudinally polarized vortices with arbitrary topological charges in terahertz (THz) frequencies using a silicon ring resonator integrated with a second-order diffraction grating. The implemented configuration enables flexible topological charge manipulation in longitudinally polarized electric fields through the excitation of quasi-transverse-magnetic (TM) waveguide modes with different frequencies. By employing a terahertz near-field measurement system, the spatial intensity patterns and phase characteristics of emitted waves are quantitatively analyzed via a precision probe. This strategy shows promising potential for applications in particle manipulation techniques and advanced imaging technologies.

Keywords: resonator; longitudinal polarization; orbital angular momentum; terahertz

1. Introduction

In the realm of vortex beams, photons carry two types of angular momentum: spin angular momentum (SAM) and orbital angular momentum (OAM). The SAM represents the polarization state of light, whereas OAM signifies the spiral phase distribution [1]. With the helical phase of vortex beams characterized as $\exp(il\phi)$, each photon carries an OAM of $l\hbar$. This imparts a doughnut-shaped intensity profile owing to the phase uncertainty at the vortex beam's center. The ability of vortex beams has sparked significant research interest, finding extensive applications in optical communication [2], optical tweezers [3], super-resolution imaging [4], and quantum information coding [5]. The generation of vortex beams is often accomplished through spiral phase plates [6], holography [7], and metasurfaces [8]. When a paraxial beam undergoes tight focusing, a longitudinal electric field component forms within the focal region [9,10]. A radially polarized field focused by a high numerical aperture (NA) lens yields a longitudinal electric field in the focal area, whilst azimuthal polarization induces a powerful magnetic field on the optical axis [11,12]. Other intriguing optical phenomena, like longitudinal polarization optical needles [13] and longitudinal vortex knots and links [14], have also been reported. These longitudinal electric fields offer promising applications in particle acceleration [15,16] and harmonic generation [17], among others. Recent advances utilizing integrated microring resonators featuring gratings have been employed to generate vortex light [18,19]. Despite significant progress in understanding transverse electric field radiation, the longitudinal counterpart remains underexplored, constrained by excitation control challenges and measurement constraints.

In this work, we propose a compact integrated THz resonator incorporating a grating structure to generate longitudinal electric field vortices through quasi-TM mode excitation. The method redirects whispering gallery modes (WGMs), which are typically confined within the ring resonator, out into free space via azimuthally periodic refractive index modulation. This mechanism produces electric field vortex beams with helical phase fronts, providing fresh insights into photonic interactions and physical processes. Unlike the previous design schemes mainly based on transverse electric field vortex light, our design enables a longitudinally polarized vortex electric field through the excitation of TM waveguide modes and topological charge control at varying operational frequencies. By using a terahertz near-field scanning system with subwavelength resolution probes, this method quantitatively maps both the spatial intensity distribution and phase profile of the radiation. This platform demonstrates promising applicability in microscale particle manipulation techniques and advanced imaging technologies.

2. Principles and Methods

Figure 1a provides an illustrative representation of the designed longitudinal electric field vortex beam emitter on a silicon platform. The ring resonator is formed by a ridge-type silicon waveguide that is 450 μm wide and 400 μm high and placed over a 130 μm slab. The radius of the resonator is 6.9 mm, while 72 grating elements are positioned along the inner sides of the ring waveguide. The grating period is approximately 582.44 μm. The duty cycle is 0.45. There is a 100 μm gap separating the bus waveguide from the ring resonator. The ridge waveguide structure features uniform cross-sectional dimensions of 400 μm (height) × 400 μm (width). The fabricated device's optical microscope images are presented in Figure 1b. The device was fabricated on a 4-inch double-side polished intrinsic silicon wafer (resistivity: 20 kΩ·cm) through a lithographic process. First, a 1.3 μm thick SiO_2 layer was deposited on the silicon wafer via plasma-enhanced chemical vapor deposition (PECVD). Subsequently, the grating and ring resonator patterns were defined on the SiO_2 hard mask using reactive ion etching (RIE). The silicon substrate was then structured through inductively coupled plasma (ICP) etching. Finally, the processed wafer was diced into discrete devices for characterization.

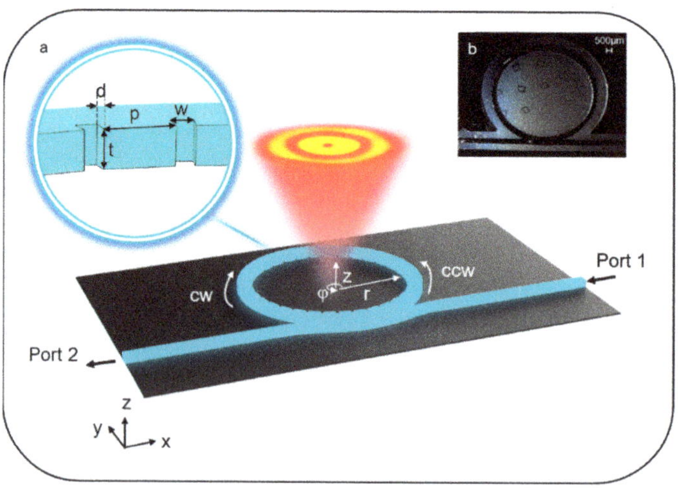

Figure 1. An illustration of the concepts. (**a**) The schematic of the designed longitudinal electric field vortex beam emitter. (**b**) Optical microscope images of the fabricated device.

In this configuration, a TM mode wave—where the dominant electric field component is oriented along the z-direction—is evanescently coupled from the bus waveguide into the ring resonator, stimulating WGMs. Angular gratings, functioning similarly to second-order gratings in straight waveguide input/output couplers, scatter the guided wave through their periodic elements. A portion of the scattered energy undergoes phase-matched interference along a specific azimuthal direction φ, generating a tilted planar wavefront at angle φ. Critically, when such a grating-integrated waveguide is bent into a ring configuration to support WGMs, Huygens' principle dictates that the radiated wavefront transitions from planar tilting to helical propagation, which is indicative of OAM-carrying beams. This transformation is physically implemented through azimuthally periodic subwavelength grating elements etched along the resonator circumference, which impose a controlled phase modulation. WGMs represent angular momentum eigenstates with discrete azimuthal propagation constants $\nu_{WGM} = \beta_p R = p$, arising from the self-consistent phase requirement for resonance under the periodic boundary condition. The integer p denotes the azimuthal mode number, where R is the effective radius of the WGM, and β_p is the propagation constants at R. The coupling between guided WGMs and radiation modes can be described by the coupled mode theory (CMT), where periodic dielectric constant variations act as perturbations coupling the unperturbed eigenmodes. Equation (1) presents the angular phase matching condition with g representing the diffraction order [18]:

$$\nu_{rad} = \nu_{WGM} - gq \qquad g = \pm 1, \pm 2, \ldots \tag{1}$$

where q denotes the number of grating periods within the resonator circumference. The implementation of a second-order grating on the WGM resonator enables the second-order diffraction to reflect light into counter-propagating waveguide modes, while the first-order diffraction (g = 1) facilitates outcoupling at a 90° angle, generating a surface-normal propagating field. Hence, when light is injected from the right input port (Port 1), it excites clockwise (CW) propagating WGMs. The topological charge (l_{TC}) of the emitted OAM beam in the $\mathbf{E_z}$ component can be derived from Equation (2) [18]:

$$l_{TC} = \pm(p - q) \tag{2}$$

Due to the z-polarized nature of the quasi-TM mode in the bus waveguide, the radiated vortex beam exhibits a dominant energy concentration in the longitudinal electric field component $\mathbf{E_z}$, which is selectively detected using a THz near-field probe optimized for longitudinal polarization sensitivity.

3. Results

To characterize the device's performance, we performed numerical simulations of its radiation electric field distribution using a commercial finite-difference time-domain (FDTD) solver (Lumerical Solutions, Inc.). The simulation setup is configured as follows: perfectly matched layer (PML) boundary conditions are applied; the mesh is dynamically generated using an auto-non-uniform algorithm; the absolute global minimum mesh step size (a lower bound for localized refinement) is set to 0.00025 μm; and a TM-polarized waveguide source is used for excitation. Figure 2a shows the z-polarization simulated electric field distribution of the quasi-TM mode in the bus waveguide. Figure 2b presents the simulated quasi-TM mode transmission spectrum, showing pronounced WGM resonances indexed by their topological charge l_{TC}. The transmission spectrum represents the ratio of the output power at port 2 to the input power at port 1 in the bus waveguide. The topological charge l_{TC} of emitted OAM beams follows Equation (1), establishing a frequency-dependent OAM control as illustrated by spectral markers in Figure 2b: at resonance frequencies of 173.8 GHz ($l_{TC} = -3$), 175.45 GHz ($l_{TC} = -2$), 177.2 GHz ($l_{TC} = -1$), and 178.65 GHz ($l_{TC} = 0$). Notably, the extinction ratio (ER) of different topological charges exhibits some fluctuations, which may stem from computational precision limitations and the inherent frequency dependence of the grating radiation efficiency and bus–ring coupling coefficient. The free spectral range (FSR) is 1.65 GHz. Calculated quality factors (Q) exhibit mode-specific variations, revealing distinct modal loss profiles: 579.33 ($l_{TC} = -3$), 701.80 ($l_{TC} = -2$), 738.33 ($l_{TC} = -1$), and 425.36 ($l_{TC} = 0$).

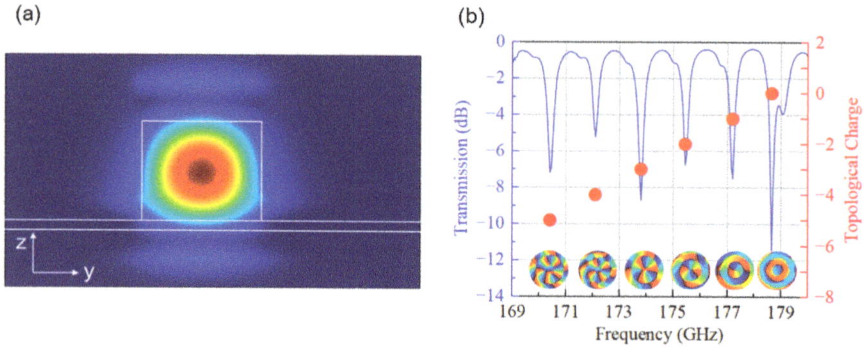

Figure 2. (**a**) The simulated $\mathbf{E_z}$ electric field distribution of the quasi-TM mode in the bus waveguide. (**b**) The simulated transmission spectrum of the THz vortex emitter for the TM mode. Note that the red circular annotation indicates the topological charge l_{TC}, and the corresponding phase result is marked below it.

Figure 3 illustrates the simulated intensity and phase distributions ($\mathbf{E_z}$) of the radiated vortex beam, alongside the mode field distribution across the cross-section of the grating-integrated ring resonator, corresponding to topological charges $l_{TC} = -3, -2, -1$, and 0. Their electric field intensities ($\mathbf{E_z}$) are normalized according to their respective maximum values. Furthermore, the mode field distribution within the grating structure demonstrates that each topological charge (l_{TC}) corresponds to a distinct resonant excitation frequency of the WGM. The characteristic doughnut-shaped intensity profiles and helical phase discontinuities validate the formation of longitudinally polarized optical vortices. The $l_{TC} = 0$ configuration displays a Gaussian-like intensity pattern with negligible OAM, acting as a null-reference for vortex verification.

The near-field experimental test system consists of the R&S ZNA26 vector network analyzer (VNA), a two-dimensional motor platform, two WR-5.1 waveguides, and two WR-5.1 140–220 GHz VNA frequency extenders, as shown in Figure 4. A WR-5.1 metallic waveguide is connected to the frequency extenders. The metallic waveguide couples vertically polarized waves into the bus waveguide via position alignment, exciting its quasi-TM mode. Energy from the WR-5.1 waveguide propagates through the bus waveguide and evanescently couples to the ring resonator. Then, it is radiated via the integrated grating, enabling OAM beam excitation. The receiver is composed of a WR-5.1 metal waveguide and a waveguide coaxial conversion probe. The probe is located at approximately one wavelength from the chip and is perpendicular to the device. The near-field probe can be used to detect z-polarized THz waves. The movement of the probe is controlled by a two-dimensional motor to achieve planar scanning of 5×5 mm^2 to obtain the THz intensity and phase distribution.

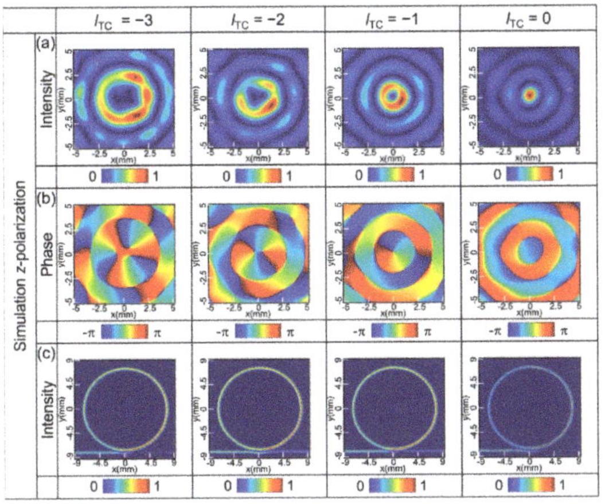

Figure 3. (**a**,**b**) The simulated near-field distributions and (**c**) mode field distribution within the grating structure of the z-polarization components for our device with the corresponding l_{TC}. Note that $l_{TC} = p - q$ is the topological charge, where p is the azimuthal order of WGM, and q is the number of grating elements.

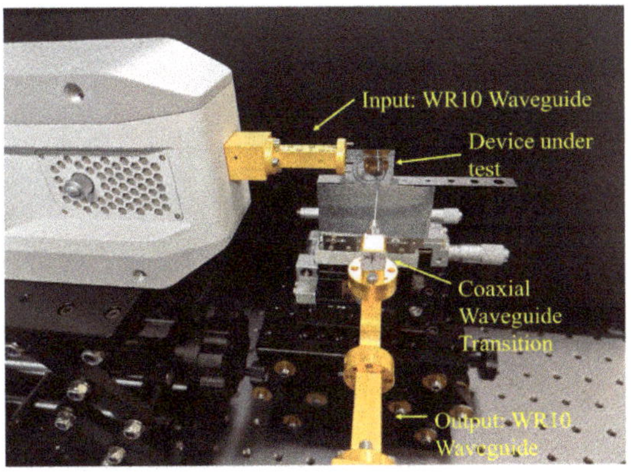

Figure 4. Experimental setup for device near-field characterization.

The transmission spectrum of the device was measured for the TM polarization using the method detailed in our prior work [20,21]. In a rectangular metallic waveguide, the electric field polarization aligns parallel to the shorter sidewall. We selectively excite quasi-TE or quasi-TM modes by orthogonally orienting the frequency extender's feed structure relative to the waveguide cross-section. Two such waveguides are positioned at opposite ends of the device's straight waveguide section, serving as THz wave input and output ports. A high-precision displacement stage actively optimizes the waveguide–device alignment, ensuring maximum coupling efficiency through real-time transmission power monitoring. The measured transmission spectrum is shown in Figure 5a. By comparing the experimental transmission spectrum with the simulation results, it can be observed that the FSR of the experiment is 0.038 GHz bigger than the simulation result, which is 1.688 GHz. The experimental result of the resonant frequency corresponding to the zero-order topological charge has shifted by 0.56 GHz in the high-frequency direction compared with the simulation result. To generate the transmission spectral response, the frequency was swept from 170 GHz to 180 GHz in 0.01 GHz increments. At resonance frequencies of 174.17, 175.77, 177.44, and 179.22 GHz, the device exhibits l_{TC} values of −3, −2, −1, and 0, respectively. The near-field distribution was characterized via scanning using a subwavelength-resolution probe, displaced across the transverse plane (x-y), enabling the acquisition of the amplitude and phase profiles presented in Figure 5b. The scanning plane encompassed an area of 5 × 5 mm^2, effectively covering the primary radiation region of the device. Notably, the measured x-o-y plane displayed distinctive doughnut-shaped intensity distributions, while the phase distributions exhibited OAM patterns. Although slight variations in intensity and inhomogeneity were observed, the qualitative agreement between the measured intensity and phase distributions and the simulation results indicated the successful generation of vortex beams.

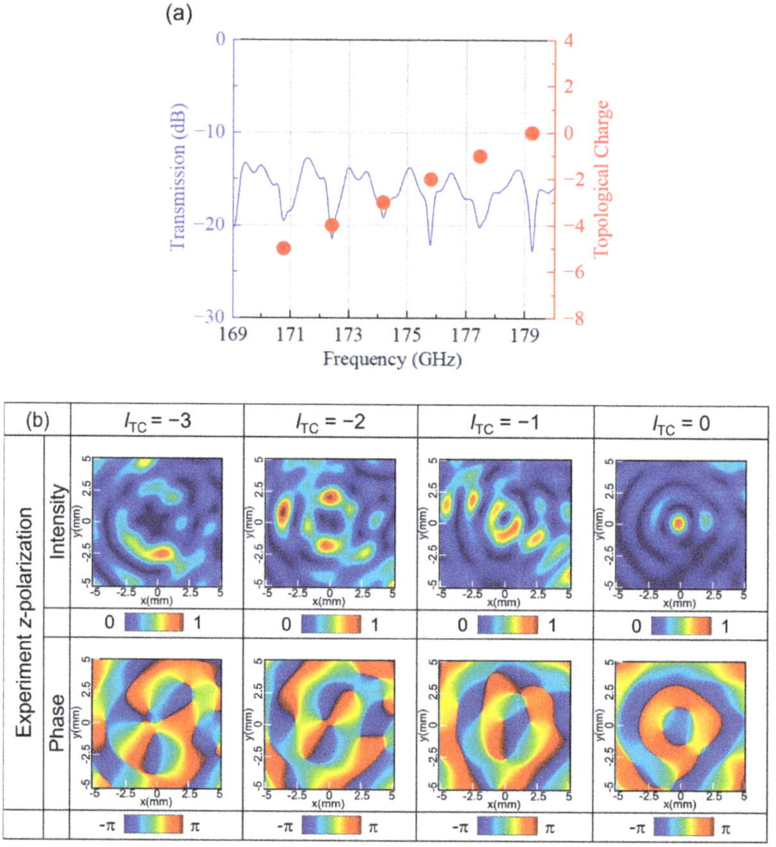

Figure 5. (**a**) The experimental transmission spectrum; the red circular annotation indicates the topological charge l_{TC}. (**b**) Measured near-field distributions of the z-polarization components for our device with the corresponding l_{TC}.

4. Discussion

The observed radiation intensity nonuniformity of the simulations and experiments (Figures 3 and 5) stem from the uniform grating structure. The designed grating exhibits a strong and consistent radiation efficiency across its entire azimuthal span. However, this leads to significant energy emissions from the input region as light initially enters the resonator. Consequently, the energy available for the subsequent propagation diminishes progressively, resulting in the gradual attenuation of the radiation intensity along the resonator's circumference. Experimental results further amplify this nonuniformity due to fabrication errors and material inhomogeneity. To address this, future designs could incorporate azimuthally chirped gratings to spatially modulate the radiation efficiency. By intentionally reducing the grating coupling strength near the input region and gradually increasing it along the propagation path, such gratings can mitigate early-stage energy radiation and enhance uniformity. The degraded circular symmetry in experimental results (Figure 5) arises from two factors: (1) stray electromagnetic waves caused by waveguide coupling leakage which interfere with the target signal and (2) the finite spatial resolution of the near-field probe, which distorts azimuthal phase profiles at high topological charges. Despite these discrepancies, key qualitative features are consistent across simulations and experiments. Both confirm helical phase structures for OAM charges (l_{TC} = −3, −2, −1, 0) and intensity minima at the vortex core (Figure 3 vs. Figure 5), validating the phase singularity mechanism.

Furthermore, a comparison between the experimental transmission spectrum and simulation results reveals a 0.038 GHz increase in the FSR (1.688 GHz measured vs. 1.650 GHz simulated). This discrepancy primarily arises from fabrication-induced deviations in the group index and ring circumference of the grating-integrated resonator, caused by the material refractive index and structural variations. Additionally, refractive index variations induce resonant frequency shifts. For instance, the resonant position corresponding to the zero-order topological charge exhibits a 0.56 GHz frequency shift between the experiment and simulation.

In addition, we used a contour integral to calculate the mode purity of different topological charges under analytical, simulated, and experimental conditions. The results are summarized in Figure 6a–d. Although the primary modes remain consistent across analytical, simulated, and experimental cases, the simulated mode purity slightly decreases compared to analytical predictions. Experimental results exhibit further purity degradation (e.g., 67.4% reduction at l_{TC} = −3) caused by stray light interference and measurement uncertainties. Future improvements in the device design (e.g., optimized grating coupling coefficients) and advanced characterization techniques are expected to enhance the purity performance.

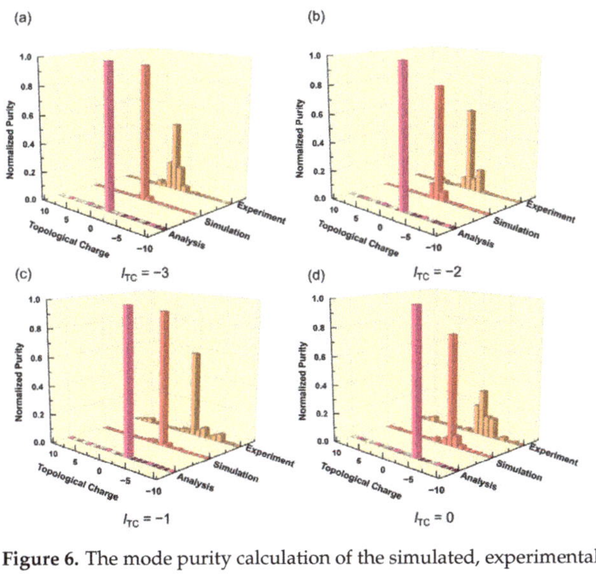

Figure 6. The mode purity calculation of the simulated, experimental, and analysis OAM modes for **(a)** l_{TC} = −3, **(b)** l_{TC} = −2, **(c)** l_{TC} = −1, and **(d)** l_{TC} = 0.

5. Conclusions

In summary, we have developed and demonstrated an integrated longitudinal polarization vortex emitter operating in the THz frequency range. This emitter allows for the tuning of topological charges via changing frequencies. Comprehensive numerical simulations and near-field measurements have been conducted to validate this approach and investigate the electric field and phase properties of the emitter. In the experiment, vortex beams with topological charges of -3, -2, -1, and 0 were successfully generated. These results highlight the promising potential of THz-integrated longitudinal polarization vortex beam chips in various applications, including the sensing and advanced imaging technologies.

Author Contributions: J.X. initiated the idea. M.J. and T.W. conducted the numerical simulations. M.J. and T.W. performed the measurements. M.J., T.W. and J.X. prepared the manuscript. All authors have read and agreed to the published version of the manuscript.

Funding: This research was funded by the National Natural Science Foundation of China grant number 62275155 and the Shanghai Rising-Star Program grant number 24QA2703200.

Institutional Review Board Statement: Not applicable.

Informed Consent Statement: Not applicable.

Data Availability Statement: The data that support the findings of this study are available from the corresponding author upon reasonable request.

Conflicts of Interest: The authors declare no competing interests.

References

1. Allen, L.; Barnett, S.M.; Padgett, M.J. *Optical Angular Momentum*; CRC Press: Boca Raton, FL, USA, 2003.
2. Jian, W. Advances in communications using optical vortices. *Photon. Res.* **2016**, *4*, 14–28.
3. Pesce, G.; Jones, P.H.; Maragò, O.M.; Giovanni, V. Optical tweezers: Theory and practice. *Eur. Phys. J. Plus* **2020**, *135*, 949. [CrossRef]
4. Jungmann, R.; Avendaño, M.; Dai, M.; Johannes, B.W.; Sarit, S.A.; Zachary, F.; Avital, R.; Peng, Y. Quantitative super-resolution imaging with qPAINT. *Nat. Methods* **2016**, *13*, 439–442. [CrossRef] [PubMed]
5. Bennett, C.; DiVincenzo, D. Quantum information and computation. *Nature* **2000**, *404*, 247–255. [CrossRef] [PubMed]
6. Ruffato, G.; Massari, M.; Romanato, F. Generation of high-order Laguerre–Gaussian modes by means of spiral phase plates. *Opt. Lett.* **2014**, *39*, 5094–5097. [CrossRef] [PubMed]
7. Wolfgang, O.; Ahmad, F.; Peng, G.; Klaus, K.; Dinesh, N.; Giancarlo, P.; Alok Kumar, S.; Mitsuo, T.; Marc, W. Recent advances in digital holography. *Appl. Opt.* **2014**, *53*, 44–63.
8. Zhang, Y.; Liu, W.; Gao, J.; Yang, X. Generating focused 3D perfect vortex beams by plasmonic metasurfaces. *Adv. Opt. Mater.* **2018**, *6*, 1701228. [CrossRef]
9. Youngworth, K.S.; Brown, T.G. Focusing of high numerical aperture cylindrical-vector beams. *Opt. Express.* **2000**, *7*, 77–87. [CrossRef] [PubMed]
10. Dorn, R.; Quabis, S.; Leuchs, G. Sharper Focus for a Radially Polarized Light Beam. *Phys. Rev. Lett.* **2003**, *91*, 233901. [CrossRef] [PubMed]
11. Quabis, S.; Dorn, R.; Eberler, M.; Glockl, O.; Leuchs, G. Focusing light to a tighter spot. *Opt. Commun.* **2000**, *179*, 1. [CrossRef]
12. Zurita-Sánchez, J.R.; Novotny, L. Multipolar interband absorption in a semiconductor quantum dot. II. Magnetic dipole enhancement. *J. Opt. Soc. Am. B* **2002**, *19*, 2722. [CrossRef]
13. Wang, H.; Shi, L.; Lukyanchuk, B.; Sheppard, C.; Chong, C.T. Creation of a needle of longitudinally polarized light in vacuum using binary optics. *Nat. Photonics* **2008**, *2*, 501. [CrossRef]
14. Maucher, F.; Skupin, S.; Gardiner, A.S.; Hughes, G.I. Creating Complex Optical Longitudinal Polarization Structures. *Phys. Rev. Lett.* **2018**, *120*, 163903. [CrossRef] [PubMed]
15. Kimura, D.W.; Kim, H.G.; Romea, D.R.; Steinhauer, C.L.; Pogorelsky, V.I.; Kusche, P.K.; Fernow, C.R.; Wang, X.; Liu, Y. Laser Acceleration of Relativistic Electrons Using the Inverse Cherenkov Effect. *Phys. Rev. Lett.* **1995**, *74*, 546. [CrossRef] [PubMed]
16. Novotny, L.; Beversluis, M.R.; Youngworth, K.S.; Brown, T.G. Longitudinal Field Modes Probed by Single Molecules. *Phys. Rev. Lett.* **2001**, *86*, 5251. [CrossRef] [PubMed]
17. Biss, P.D.; Brown, G.T. Polarization-vortex-driven second-harmonic generation. *Opt. Lett.* **2003**, *28*, 923. [CrossRef] [PubMed]

18. Cai, X.; Wang, J.; Strain, M.J.; Cai, X.; Wang, J.; Strain, M.J.; Johnson-Morris, B.; Zhu, J.; Sorel, M.; O'Brien, L.J.; et al. Integrated compact optical vortex beam emitters. *Science* **2012**, *338*, 363–366. [CrossRef] [PubMed]
19. Xie, J.; Qian, J.; Wang, T.; Zhou, L.; Zang, X.; Chen, L.; Zhu, Y.; Zhuang, S. Integrated terahertz vortex beam emitter for rotating target detection. *Adv. Photonics* **2023**, *5*, 066002. [CrossRef]
20. Xie, J.; Zhu, X.; Zang, X.; Cheng, Q.; Chen, L.; Zhu, Y. Terahertz integrated device: High-Q silicon dielectric resonators. *Opt. Mater. Express.* **2018**, *8*, 50–58. [CrossRef]
21. Xie, J.; Zhu, X.; Zhang, H.; Zang, X.; Chen, L.; Alexey, V.B.; Alexander, P.S.; Zhu, Y. Terahertz-frequency temporal differentiator enabled by a high-Q resonator. *Opt. Express.* **2020**, *28*, 7898–7905. [CrossRef] [PubMed]

photonics

MDPI

Article

Grooved High-Reflective Films for Ultraviolet Emission Enhancement

Hengrui Zhang, Zhanhua Huang and Lin Zhang *

State Key Laboratory of Precision Measuring Technology and Instruments, Key Laboratory of Opto-Electronic Information Technology of Ministry of Education, Tianjin Key Laboratory of Integrated Opto-Electronics Technologies and Devices, School of Precision Instruments and Opto-Electronics Engineering, Tianjin University, Tianjin 300072, China; hengruizhang@tju.edu.cn (H.Z.); zhanhua@tju.edu.cn (Z.H.)

* Correspondence: lin_zhang@tju.edu.cn

Abstract

Conventional ultraviolet microplasma sources typically lack a back-reflection structure, resulting in radiative power loss from the backside. To enhance the emission efficiency of ultraviolet microplasma devices around 220 nm, we propose a multilayer reflective coating composed of alternating high- and low-refractive-index layers of Al_2O_3 and SiO_2, within a V-shaped groove. Key structural parameters, including the number of alternating film layer pairs, groove width, and light source position, are investigated to show their effects on ultraviolet reflection characteristics. The results show that reducing the groove width greatly enhances light reflection. When the groove width is 6.5 μm, the device exhibits a reflection efficiency of 47.82% and power enhancement of 91.66%, representing improvements of 2.5-fold and 4.2-fold, respectively, compared to non-optimized cases. Device performance is also influenced by the offset of the light source, which is more sensitive along the horizontal direction. This study provides a practical solution for developing high-efficiency ultraviolet emission devices.

Keywords: ultraviolet high-reflective films; groove-shaped device; emission enhancement

1. Introduction

Microplasma, a novel form of electrical discharge, have attracted considerable attention due to their high energy density, elevated plasma density, and exceptional spatiotemporal stability at the microscale [1–3]. Compared to conventional plasmas, microplasma can be generated and sustained at or below atmospheric pressure with appropriate design, enabling the formation of micrometer-scale discharges while significantly reducing system complexity and cost. Microplasma has been extensively studied and applied across various fields, including ultraviolet (UV) light sources [4–7], beam steering [8,9], microchemical analysis systems [10], biomedicine [11,12], and material surface modification [13], due to their excellent controllability and broad applications. Ultraviolet light sources operate within a frequency range of 7.5×10^{14} Hz to 3×10^{16} Hz, corresponding to vacuum wavelengths of 10–400 nm. Ultraviolet radiation is classified into four distinct spectral bands based on wavelength range and biological effects: UV-A (315–400 nm), UV-B (280–315 nm), UV-C (100–280 nm), and vacuum ultraviolet (10–100 nm) [14]. Conventional UV-C sources primarily employ mercury vapor lamps that emit light at a wavelength of 254 nm. Preliminary studies suggest that ultraviolet light at 222 nm is equally effective in inactivating viruses and bacteria while offering improved safety for human exposure [15,16].

In recent years, microplasma arrays fabricated using microfabrication techniques have developed rapidly [17–24], in which the spatial confinement and coordinated control of plasma, enabled by the periodic arrangement of micron-scale discharge units, are needed. For example, in 1997, Eden's research group at the University of Illinois, USA, fabricated a single-cavity plasma device with a cylindrical structure on a silicon-based substrate using ultrasonic machining [17]. The device had a diameter of 200–400 μm and a depth of 0.5–5 mm. When filled with a gas mixture of 50 Torr Xe and 1 Torr I_2, and operated at an input power of 500 mW, it produced an output power of 270 mW, measured at a distance of 1.4 cm from the device. In 2005, they proposed an inverted-pyramidal microcavity plasma discharge device composed of a 200 × 200 array, with each microcavity in a size of 50 × 50 μm^2 [19]. When driven by a 20 kHz sinusoidal voltage, the device exhibited a luminous efficacy of 7.2 ± 0.6 lm/W and a brightness of 525 ± 75 cd/m^2. In 2017, they engineered a microplasma array consisting of at least two interleaved microcavities [23]. The configuration, which employed two fused silica plates as front and rear windows, produced a planar lamp in the size of 10 × 10 cm^2, with an effective area of 80 cm^2 and a thickness of less than 6 mm. The device is operated in a 54% Xe/Ne gas mixture at 550 Torr, with an electro-optical conversion efficiency exceeding 20%. However, due to the absence of a back-reflection structure, over 40% of the radiated power is lost through the backside. This emphasizes the importance of adding reflective multilayers. In 2020, Dzikowski et al. from Ruhr University Bochum, Germany, introduced a modular metallic grating microcavity plasmonic array [24]. The array consisted of several hundred uniformly distributed apertures, each with a diameter of 150 μm, and employed magnetic forces to maintain the structural integrity of the assembly. Current research on ultraviolet light source devices primarily focuses on the design of microcavity array architectures and the enhancement of emission intensity. In contrast, systematic studies aimed at efficient UV light collection, output amplification, and improved energy utilization remain relatively limited. Consequently, effectively enhancing the output of ultraviolet light at specific wavelengths is essential for improving the overall performance of plasma emission systems.

The high resolution and repeatability of advanced lithography and etching techniques in micro–nano fabrication processes provide a solid foundation for the integration and miniaturization of multifunctional devices. Reflective thin films for ultraviolet applications can be highly effective via multilayer interference. In 2014, Sun et al. at the Shanghai Institute of Optics and Fine Mechanics designed an Al_2O_3/AlF_3 reflective coating [25]. A reflectance of 98.1% was achieved in the 180–200 nm spectral range. In 2017, Narukage et al. from the National Astronomical Observatory of Japan designed a MgF_2/LaF_3 reflective filter [26], achieving a peak reflectance of 57% at 121.6 nm. However, the stability of these reflective coatings is compromised by the degradation and contamination of fluoride materials during storage. Although a ZrO_2/Al_2O_3 bandpass filter was reported in 2023 for suppressing long-wavelength emissions in the 240–270 nm range from KrCl* excimer lamps operating at 222 nm [27], it is even more noteworthy that recent advances since 2020 in fields, such as metasurface reflectors, temporal modulation, and 3D photonic crystals, would inspire great progresses in device innovations [28–30].

In this paper, we study a groove-shaped ultraviolet high-reflective film based on COMS-compatible fabrication, to enhance the luminous efficiency and upward output of UV microplasma devices. Structural design and optimization are conducted. On one hand, the groove width plays a crucial role in determining the reflective performance of the groove-shaped device. When the groove width is set to 6.5 μm, the device achieves optimal reflective performance, with a reflectance efficiency of up to 47.82% and a power enhancement of 91.66%, representing 2.5-fold and 4.2-fold improvements, respectively, compared to the non-optimized case. On the other hand, displacement of the light source has a pronounced effect on device's performance. Horizontal displacement causes substantial performance variations, while vertical displacement exhibits spatial selectivity.

2. Structural Design

We propose a groove-shaped device to build a microcavity for microplasma, while a high-reflectivity multilayer coating is used to enhance the upward light extraction efficiency of ultraviolet emission systems, with its cross-sectional view shown in Figure 1a. Here, W denotes the groove opening width, and the apex angle is 68°, which is determined by wet etching of silicon. We design the multilayer reflective films composed of alternating layers of Al_2O_3 and SiO_2, centered at a wavelength of 220 nm. Al_2O_3 serves as the high-refractive-index material, while SiO_2 functions as the low-refractive-index material. Silicon wafers are selected as the substrate, onto which Al_2O_3 and SiO_2 can be alternately formed using the atomic layer deposition process. The final structure includes a terminating layer of Al_2O_3 and can be represented as $Si/(HL)^mH/Air$, where H denotes Al_2O_3, L denotes SiO_2, m is the number of alternating periods, and the total number of layers is $2m + 1$. The optical thickness of each film layer is set to $\lambda/4n$, where λ is the designed central wavelength in vacuum, and n is the refractive index of the corresponding film material. Based on the effective interface method [31] for multilayer films, the reflectance of the structure under normal incidence and at the central wavelength can be obtained through a layer-by-layer calculation as follows:

$$\rho_\lambda = \left[\frac{n_0 - \left(\frac{n_H}{n_L}\right)^{2m} \frac{n_H^2}{n_G}}{n_0 + \left(\frac{n_H}{n_L}\right)^{2m} \frac{n_H^2}{n_G}} \right]^2, \quad (1)$$

where n_0 is the refractive index of the incident medium, typically assumed to be 1, while n_H and n_L are the refractive indices of the high- and low-index layers, respectively. A larger contrast between n_H and n_L, along with an increased total number of layers, results in higher reflectance of the multilayer film structure. According to the experimental report [32,33], the refractive indices of Al_2O_3 and SiO_2 at 220 nm are set to n_H = 1.7975 and n_L = 1.5355, respectively.

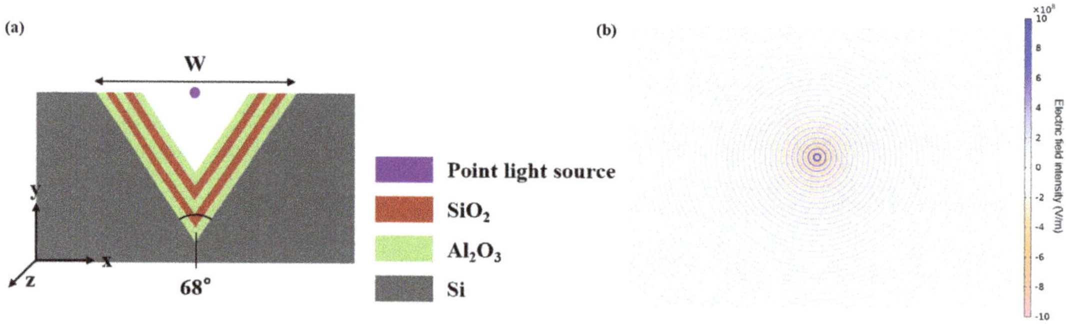

Figure 1. (**a**) Schematic cross-sectional view of the groove-shaped device. (**b**) Simulated electric field intensity distribution from a point source within the device.

To quantitatively evaluate the reflective performance, we investigate two cases: one without the coatings and another with the reflective coatings. By comparing the electric field distributions and far-field radiation patterns between these two cases, the improvement in ultraviolet light reflection provided by the groove-shaped film can be clearly observed.

The light source excited by the microplasma has its radiation characteristics approximated as those of an ideal point source. The electric field distribution shown in Figure 1b is obtained by simulating the radiation of a point source using finite-element modeling.

Specifically, the point source is approximated using the line current setting. The field distribution of the point source exhibits a standard spherical wavefront, consistent with the propagation behavior of a point source in an ideal homogeneous medium.

3. Results and Discussion

3.1. Analysis of the Influence of Alternating Film Layer Periods

We calculated the reflection spectra under normal incidence over the wavelength range of 180 to 260 nm for four different multilayer structures with alternating period numbers of $m = 5, 7, 9$, and 11, using the Transfer Matrix Method. As shown in Figure 2, the reflectance of the films increases significantly with the number of alternating periods.

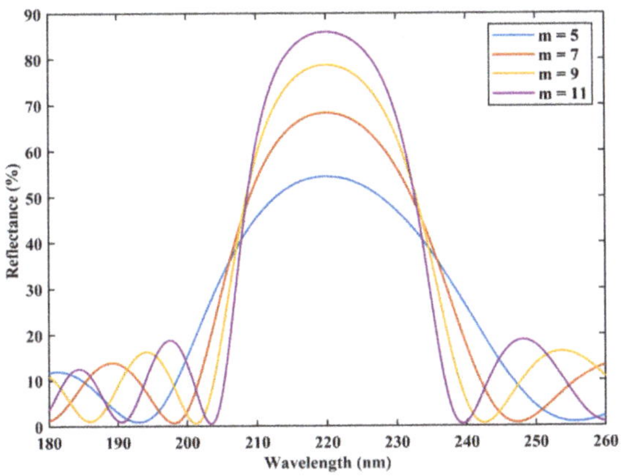

Figure 2. Reflection spectra for different numbers of alternating film layer periods.

Specifically, when $m = 5$, the reflection peak is relatively broad and smooth, with a reflectance of 54.41%. When $m = 7$, a secondary peak begins to emerge on the shorter-wavelength side, and the reflectance increases to 68.24%. As m increases further, the secondary peak structure becomes more pronounced, exhibiting multiple distinct periodic sub-peaks. When $m = 11$, the reflectance reaches 85.93%, with a full width at half maximum of 25.45 nm.

In addition to normal incidence, oblique incidence should also be considered. Figure 3 shows the variation in reflectance at different incident angles for $m = 11$. At an incident angle of 10°, the reflection spectrum exhibits only a slight shift compared to normal incidence. The central wavelength of the main peak remains close to 220 nm, with a slight shift toward the shorter wavelength region. As the incident angle increases to 30°, the main peak shifts further toward shorter wavelengths, and the peak reflectance exhibits a slight decreasing trend. This phenomenon can be attributed to the decrease in the effective optical path length within the multilayer films as the incident angle increases, causing a blue shift of the central wavelength. Additionally, the reflectance coefficients of the polarized components at the interfaces between high- and low-refractive-index materials vary with incident angle, leading to a moderate reduction in overall reflectance. When the incident angle increases to 50°, noticeable changes occur in the spectral profile. The main reflection peak shifts to around 200 nm, while a pronounced minimum appears near 220 nm. At an incident angle of 70°, the reflection spectrum shows further redistribution of peaks and valleys. The main peak shifts toward 190 nm, with the peak reflectance reduced to 70%, and a secondary minimum appears near 220 nm. This indicates that at large incident angles, the phase-matching conditions within the multilayer stack are significantly restructured for different wavelengths. Consequently, the primary high-reflectance band formed under normal or small-angle incidence may become redistributed into multiple secondary interference peaks, thereby reducing reflectance at the design wavelength. In general, the reflective behavior of the multilayer structure exhibits angular dependence. As the incident angle increases, the reflection peak shifts toward shorter wavelengths, accompanied by a noticeable reduction in peak reflectance.

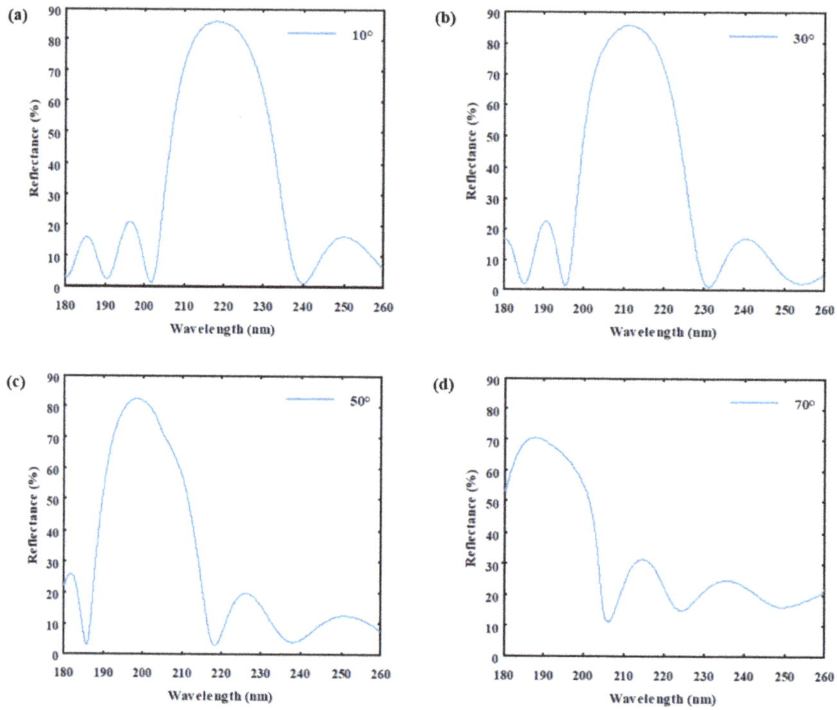

Figure 3. Reflection spectra at different incident angles for $m = 11$. (**a**) Incident angle of 10°. (**b**) Incident angle of 30°. (**c**) Incident angle of 50°. (**d**) Incident angle of 70°.

In practice, the angle of incidence on the film surface varies depending on the positions of the point light sources. Therefore, considering only normal incidence or a single oblique angle is insufficient to characterize the real optical behavior. It is necessary to evaluate the overall reflective performance of the multilayer system and further investigate its response to radiation from the light source.

3.2. Analysis of the Influence of Groove Width

We use a frequency-domain electromagnetic wave analysis tool, which was built in- house based on the finite element method to obtain the electric field distribution and far-field radiation patterns. We conduct simulations at a wavelength of 220 nm, using the corresponding refractive indices for all materials at 220 nm and applying perfectly matched layers [34] at the boundaries of our computation window. A triangular mesh is applied, with a minimum element size of 15 nm near interfaces and grooves. The near-field electromagnetic response of the device is first computed, and the far-field radiation pattern is then obtained using a far-field transformation. Each V-grooved discharge unit has a depth of about 50 μm [35] perpendicular to the x–y plane as in Figure 1a, which is >200 times the wavelength in vacuum, i.e., 220 nm, making field variation along z-direction negligible. Thus, the electro-magnetic responses of the proposed device in the z-dimension can be viewed as effectively uniform. To examine the influence of the structural dimensions of the V-shaped groove on reflective performance, we first consider a case with a groove width, W, of 10 μm as an example. The resulting electric field distribution and far-field radiation pattern are shown in Figure 4.

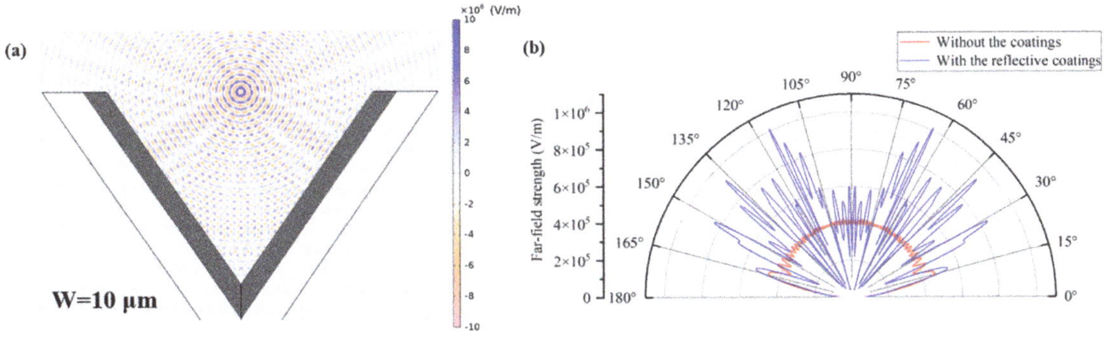

Figure 4. For W = 10 μm, (**a**) electric field intensity distribution and (**b**) far-field radiation pattern are given.

Figure 4a shows the electric field distribution. Compared to Figure 1b, multiple reflections of electromagnetic waves occur within the V-shaped groove, resulting in the formation of complex standing waves and interference fringes, which indicate local field enhancement. Due to the refractive index contrast between the high- and low-index layers of the multilayer structure, the incident electromagnetic wave undergoes successive partial reflections and transmissions at each interface. These processes cause effective interference between the reflected and transmitted waves within the microscale groove, thereby forming a stable standing wave field distribution. Figure 4b shows the far-field radiation pattern. In the case without the coatings, it exhibits a hemispherical distribution with uniformly distributed energy. In contrast, the presence of several sharp lobes in the case with the reflective coatings suggests that multilayer reflections within the groove lead to a complex far-field radiation pattern. Moreover, the magnitude of the far-field mode shows that the lobe intensity in the case with reflective coatings, which spans angular ranges of approximately 60–75° and 105–120°, is nearly twice that of the case without the coatings, further confirming that multiple reflections within the groove cause constructive interference and upward enhancement.

To quantitatively evaluate the reflective performance of the groove, we further introduce reflection efficiency and power enhancement. These are inspired by prior studies that report relative metrics such as luminescence intensity and conversion efficiency [23]. By integrating the far-field radiation patterns corresponding to the case without the coatings and the case with reflective coatings, capturing all upward-emitted components, the associated radiated powers emitted above the groove are obtained and denoted as P_1 and P_2, respectively. These quantities correspond to the output optical power above the device. The reflection efficiency can be defined as follows:

$$\eta = \frac{P_2 - P_1}{P_2}. \tag{2}$$

This ratio represents the proportional contribution of the reflective structure to the far-field radiation and reflects its effective utilization in the total radiated energy. It serves as a useful metric for evaluating the effectiveness of the proposed device in terms of reducing losses and improving energy utilization. The power enhancement can be defined as follows:

$$G = \frac{P_2 - P_1}{P_1}. \quad (3)$$

This ratio represents the percentage increase in radiation for the case with reflective coatings relative to the case without coatings, indicating the effectiveness of the groove-shaped multilayer device in enhancing overall radiation. It can be used to evaluate the ratio of the increased radiated power to the original radiated power.

The far-field radiation data are integrated, yielding the radiated powers P_1 and P_2 for the cases without and with reflective coatings, respectively. By substituting these values into Equations (1) and (2), the reflection efficiency and power enhancement for W = 10 μm are calculated to be 42.87% and 75.03%, respectively. They are calculated for different groove widths, as shown in Figure 5.

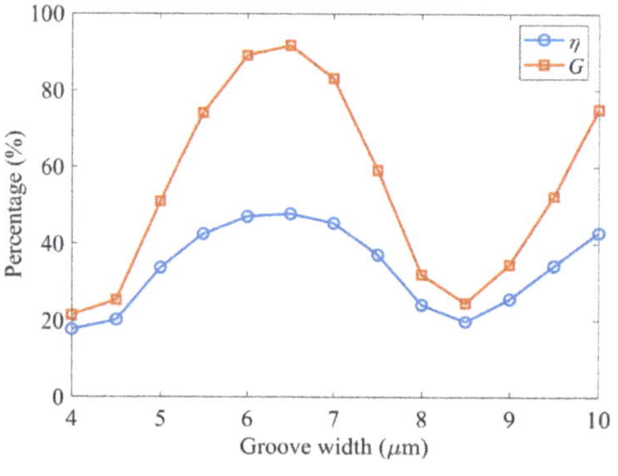

Figure 5. Effects of groove width on reflective performance.

When the groove width is reduced from 10 μm to 8.5 μm, the reflection efficiency decreases to 19.77%, and the power enhancement drops to 24.64%. The corresponding electric field intensity and far-field radiation pattern are shown in Figure 6. As shown in Figure 6a, the multilayer film within the groove continues to change the electric field. However, the interference fringes are significantly less distinct and concentrated compared to the case with W = 10 μm. The electromagnetic waves exhibit increased divergence near the groove opening. Along the reflection path, destructive interference becomes more pronounced, indicating that the optical path differences at multiple positions no longer correspond to integer multiples of the wavelength, thereby suppressing constructive interference. Meanwhile, Figure 6b shows that both the amplitudes and the number of lobes are greatly reduced. The radiation intensity in multiple directions is noticeably weakened, showing that the conditions for multipath interference are no longer satisfied in this case, leading to dispersion and attenuation of the radiated energy.

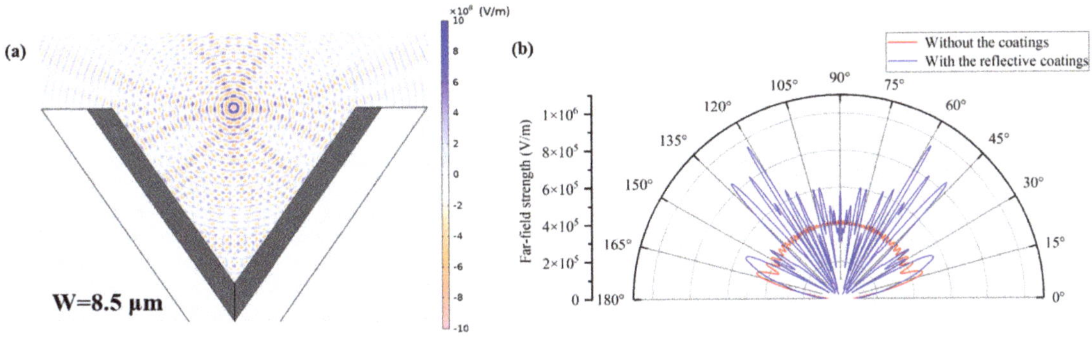

Figure 6. For W = 8.5 μm, (**a**) electric field intensity distribution and (**b**) far-field radiation pattern are given.

When the groove width is further reduced from 8.5 μm to 6.5 μm, the device exhibits optimal reflective performance, with reflection efficiency and power enhancement reaching peak values of approximately 47.82% and 91.66%, respectively. The corresponding electric field intensity and far-field radiation pattern are shown in Figure 7. As shown in Figure 7a, distinct and intense interference fringes are formed within the V-shaped groove, exhibiting strong constructive interference. The electric field is highly concentrated around the groove opening, exhibiting a high-density distribution. The wavefront displays good symmetry and radiates outward in the vertical direction, showing that the energy undergoes multiple reflections within the cavity and is effectively coupled and emitted in a stable direction.

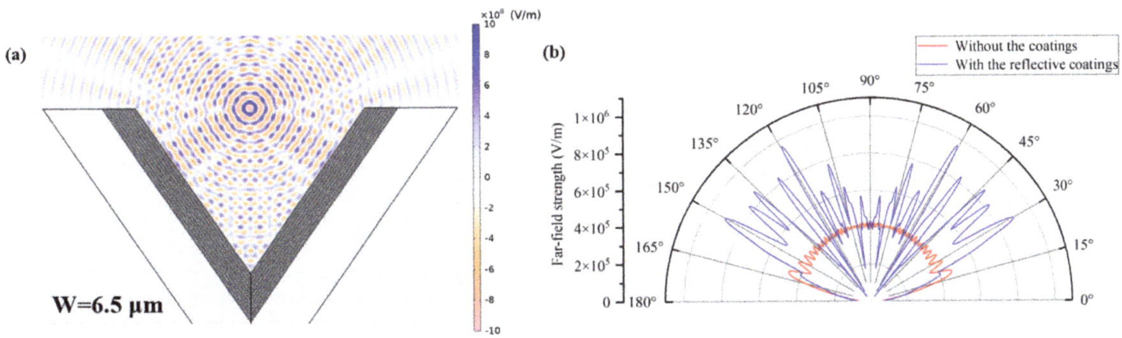

Figure 7. For W = 6.5 μm, (**a**) electric field intensity distribution and (**b**) far-field radiation pattern are given.

As shown in Figure 7b, the number of lobes increases, and the radiation intensity is significantly enhanced in multiple directions compared to Figure 4b. This phenomenon can be attributed to the enhanced phase accumulation effect resulting from multiple reflections within the cavity, which leads to more pronounced interference superposition. As a result, stronger upward radiation characteristics and a more concentrated energy distribution are observed in the far field. Furthermore, the enhancement of lobes also shows that local coherent superposition is simultaneously achieved along multiple interference paths, further improving overall radiation efficiency. Consequently, a groove width of W = 6.5 μm can be considered the optimal design parameter, as it effectively enhances both reflection efficiency and energy output.

As the groove width is further reduced to 4 μm, the reflective performance deteriorates significantly. The reflection efficiency decreases to 17.69%, and the power enhancement drops to 21.49%. The corresponding electric field intensity distribution is shown in Figure 8. The interference fringes in the V-shaped groove region appear sparse, with an overall reduction in field strength. Although some reflection still occurs within the groove, the reflected wavefronts fail to achieve phase matching, preventing significant coherent superposition. As a result, the energy reutilization efficiency within the structure is substantially reduced. This result indicates that when the groove becomes too narrow, the propagation and reflection paths of the light waves are severely constrained, ultimately leading to a substantial reduction in the structure's ability to release reflected energy. This manifests as a degradation in reflective performance.

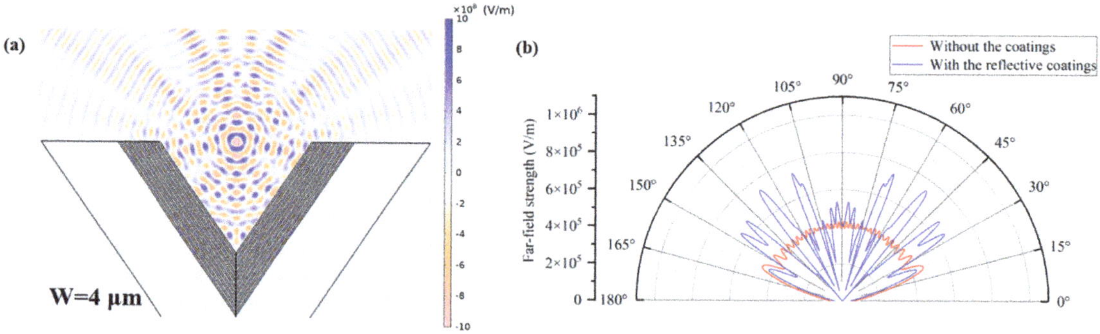

Figure 8. For W = 4 μm, (**a**) electric field intensity distribution and (**b**) far-field radiation pattern are given.

As shown in Figure 8b, although the intensity is slightly enhanced compared to the case without the coatings, the overall radiation remains weak. The far-field energy distribution exhibits a low-intensity diffusion pattern. This radiation pattern shows that under these conditions, the reflected energy within the groove cavity cannot be effectively directed outward. These characteristics indicate that when the groove width is reduced below a certain threshold, the device can no longer establish an effective upward cavity-reflection mechanism. Even though multiple reflections may still occur, the lack of constructive interference leads to dispersed energy distribution. Ultimately causing low far-field output, limited reflection efficiency, and minimal power enhancement.

Therefore, by appropriately tuning the geometric dimensions of the groove, the internal reflection paths, phase accumulation mechanisms, and interference effects of electromagnetic waves can be effectively controlled, resulting in a notable enhancement in the device's reflective performance.

3.3. Analysis of the Influence of Light Source Position

Under practical experimental conditions, the ultraviolet light source generated via microplasma is often not precisely aligned with the geometric center of the groove-shaped device due to inherent limitations in alignment accuracy. Therefore, we further investigate the effects of source offset relative to the center of the groove structure. The results obtained with a groove width of W = 10 μm and the light source positioned at the geometric center of the device serve as the reference. Figure 9 shows the electric field intensity and far-field radiation pattern resulting from a 1 μm offset of the light source along the x-axis, with the groove width and other structural parameters at constant.

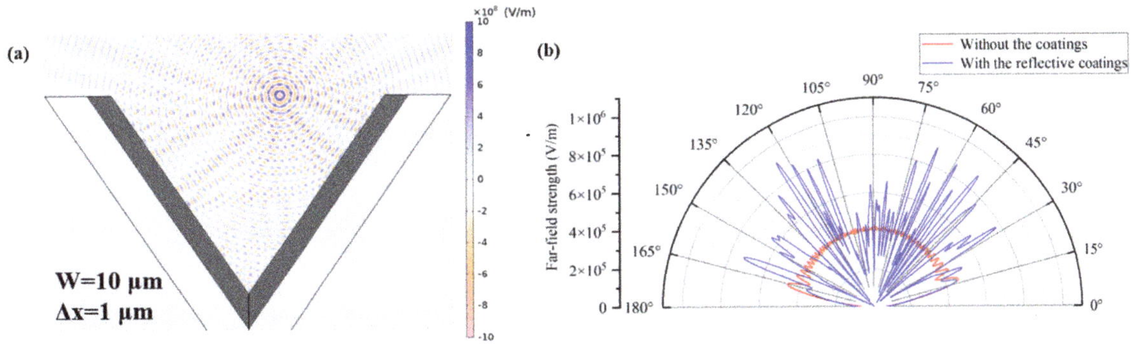

Figure 9. For a 1 μm x-axis offset of the light source, (**a**) electric field intensity distribution and (**b**) far-field radiation pattern are given.

When the light source is displaced along the x-axis, the wavefronts incident on the left and right walls of the groove experience unequal optical path lengths, leading to an initial phase difference. As shown in Figure 9a, the interference fringes on the right side of the groove become denser with significantly enhanced local field intensity, whereas those on the left side appear sparser with reduced field strength. This asymmetric field enhancement primarily arises from the accumulated phase difference caused by variations in propagation path lengths when the light source is displaced from the central position, resulting in uneven coherent superposition of the reflected waves within the groove.

The corresponding far-field radiation pattern is shown in Figure 9b. It can be observed that the radiation direction no longer exhibits the original symmetrical distribution. The asymmetry in the lobe distribution arises from the optical path difference between the two groove walls. The offset causes shorter optical paths for reflection on the nearer side, resulting in denser and more intense lobes concentrated in 45–75°. Conversely, reflection from the farther side experiences longer paths, leading to sparser lobes distributed between 105 and 135°. This phenomenon shows the complex radiation characteristics resulting from the combined effects of propagation path length differences and interference at the multilayer interfaces under source offset conditions.

The reflection efficiency and power enhancement at different offset values of the light source along the x-axis are calculated, as shown in Figure 10.

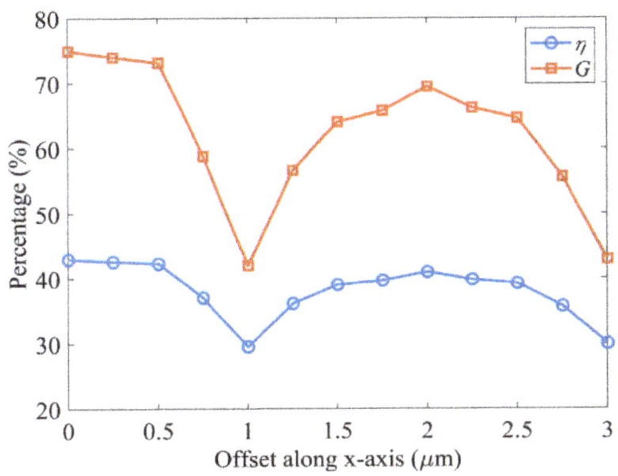

Figure 10. Effects of light source offset along the x-axis on reflective performance.

It is observed that as the light source offset increases, the reflection efficiency slightly decreases, reaching a minimum of 29.57% at an offset of 1 μm. Simultaneously, the power enhancement drops significantly to 41.99%, indicating that the structural symmetry is disrupted at this offset, leading to phase interference along the reflected light paths and subsequently reducing the far-field energy output.

When the offset is further increased to 2 μm, both metrics show partial recovery, with the reflection efficiency rising to 40.99% and the power enhancement reaching 69.47%. However, neither returns to the initial values. The corresponding electric field intensity and far-field radiation pattern are shown in Figure 11. The interference fringes of the electric field become increasingly asymmetric, with reflection paths enhanced on the side closer to the offset and weakened on the opposite side. As shown in Figure 11b, the number of lobes increases significantly on the side closer to the offset, accompanied by enhanced radiation intensity in multiple directions. This may be attributed to changes in the incident angle of light at fixed positions on the multilayer surface as the offset increases, causing improved reflective performance compared to the case with a 1 μm offset.

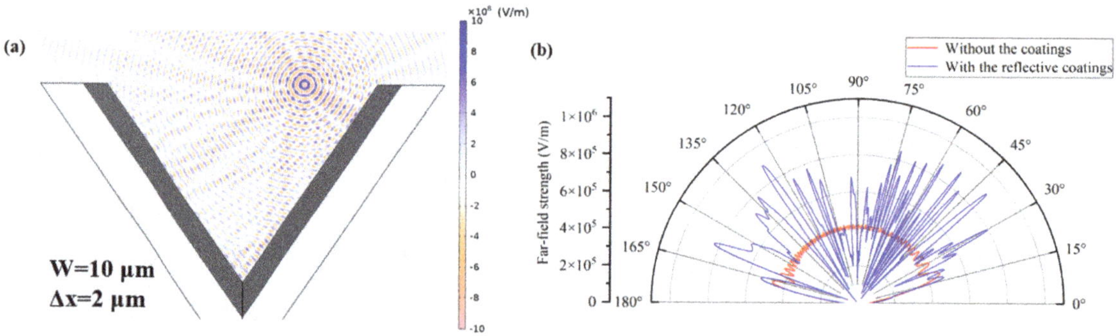

Figure 11. For a 2 μm x-axis offset of the light source, (**a**) electric field intensity distribution and (**b**) far-field radiation pattern are given.

However, as the offset continues to increase, the reflective performance drops sharply once again, indicating that the effective reflective region of the structure is spatially bounded. Beyond this range, the radiated energy can no longer be efficiently controlled. This result shows that the groove exhibits high sensitivity to the light source position, and its alignment accuracy directly affects the performance of the reflected radiation.

In addition to the x-axis offsets, we also perform simulations and analyses for light source offsets along the y-axis. When the light source is offset by 1 μm along the y-axis, the electric field intensity and far-field radiation pattern are shown in Figure 12. The light source is positioned closer to the bottom region of the groove, significantly shortening the propagation distance of the initial incident wavefront to the sidewalls and thereby reducing the internal reflection path length of the electromagnetic waves. This change in the propagation path directly causes an overall upward shift of the interference fringes inside the groove, which signifies more energy being funneled toward the groove opening. The regions of enhanced field intensity move noticeably toward the groove opening, and a strong interference-enhanced band appears near the opening region.

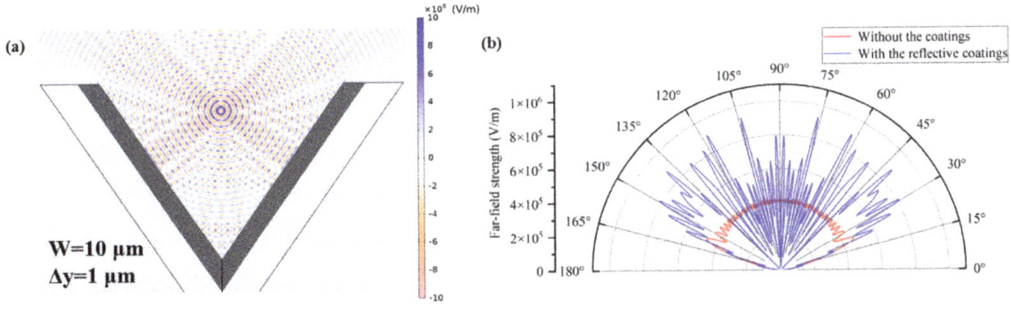

Figure 12. For a 1 μm y-axis offset of the light source, (**a**) electric field intensity distribution and (**b**) far-field radiation pattern are given.

Meanwhile, as shown in Figure 12b, when the light source is offset along the y-axis, the lobe intensity in the far-field radiation shows no significant change compared to the case where the light source is positioned at the geometric center of the groove. However, it is noteworthy that the number of lobes along the 90° direction increases significantly, accompanied by a denser spacing between lobes. This variation in lobe distribution shows that the downward offset of the light source shortens the initial reflection paths between the groove walls, thereby enhancing the coherent superposition of electromagnetic waves near the groove opening. Consequently, the radiation becomes more strongly concentrated in the direction perpendicular to the groove opening, leading to a more pronounced upward radiation pattern. The reflection efficiency at this offset is calculated to be 42.34%, with a corresponding power enhancement of 73.43%. This result indicates that the device's reflective performance undergoes only a slight reduction and remains within an acceptable range.

Figure 13 shows the variations in reflection efficiency and power enhancement under different y-axis offsets of the light source. It is observed that the structure exhibits a periodic response during the y-direction offset process, with performance peaks appearing at multiple offset positions. Specifically, when the offset reaches 0.25 μm, the reflective performance is slightly improved compared to the initial position. As the offset increases to 0.5 μm, both the reflection efficiency and power enhancement drop significantly, reaching 31.6% and 46.2%, respectively. At an offset of 1 μm, both metrics rise again, with the reflection efficiency reaching 45.81% and the power enhancement increasing to 84.55%. As the light source is further offset to 1.5 μm, the performance declines once more. However, at 2 μm, a substantial improvement is observed, with the reflection efficiency and power enhancement reaching 52.96% and 112.59%, respectively, which represent the highest values within the scanned range. Subsequently, a local peak also appears at an offset of 2.75 μm, while at 2.5 μm and 3 μm, the reflective performance exhibits varying degrees of decline.

These results show that the groove exhibits a certain degree of spatial selectivity to offsets along the y-axis. Variations in reflection paths and phase-matching conditions at different offsets jointly determine whether interference enhancement or suppression occurs within the structure. In particular, an offset of 2 μm appears to create more favorable conditions for coherent superposition, thereby enhancing reflective performance and significantly boosting power output.

According to the analysis of Figures 10 and 13, the groove is more sensitive to offsets along the x-axis, while exhibiting a certain degree of spatial selectivity along the y-axis under the condition of ensuring efficient coupling and reflection of the point source. Nevertheless, regardless of the light source position, the structure consistently maintains relatively good reflective performance within a certain range.

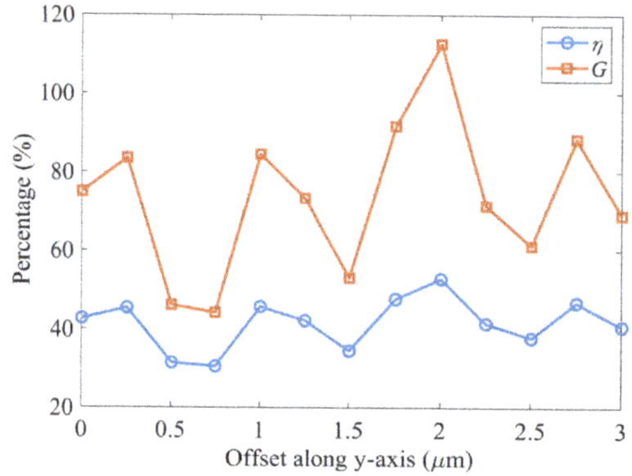

Figure 13. Effects of light source offset along the y-axis on reflective performance.

3.4. Analysis of the Influence of Multiple Light Sources

To examine real conditions more accurately, we further investigate the electromagnetic response of the coatings in a 10 μm-wide groove by arranging multiple point sources. Specifically, nine point sources are placed in a tilted square array, positioned at the center, the four corners, and the midpoints of the four sides of the square.

Figure 14 shows the corresponding electric field distributions. Without the coatings, the multiple point sources generate interference fringes, while about 34% of UV light is absorbed by the silicon substrate. In contrast, with the reflective coatings, we see enhanced standing wave patterns both inside and above the groove. Notably, previously downward UV light undergoes multiple internal reflections and is redirected upward.

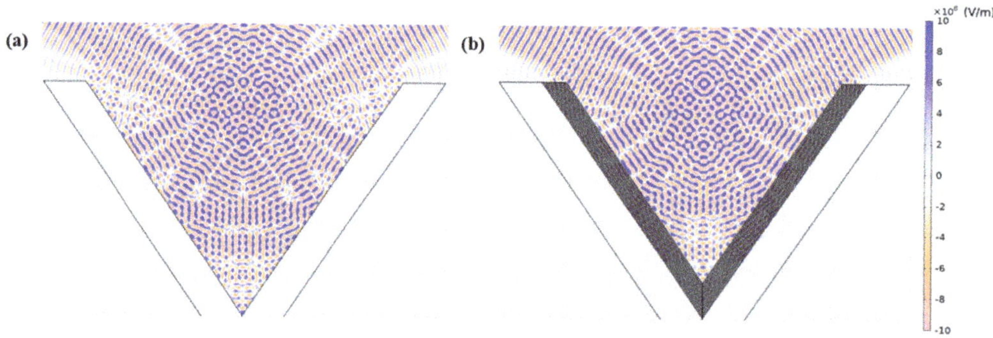

Figure 14. Electric field intensity distribution for multiple point sources. (**a**) Without the coatings. (**b**) With the reflective coatings.

The corresponding far-field radiation patterns are shown in Figure 15. Compared to the case without coatings, significantly enhanced upward radiation is found. In particular, the intensities at 45° and 135° increase by nearly threefold. This shows that the groove facilitates constructive interference within specific angular ranges, thereby enhancing upward radiation. Further calculations reveal that the reflection efficiency and power enhancement of this configuration increase to 42.35% and 73.41%, respectively. These results further confirm the effectiveness of the V-grooved reflective coatings in enhancing

UV light extraction, suppressing backward losses, and improving the overall utilization efficiency of the emitted light.

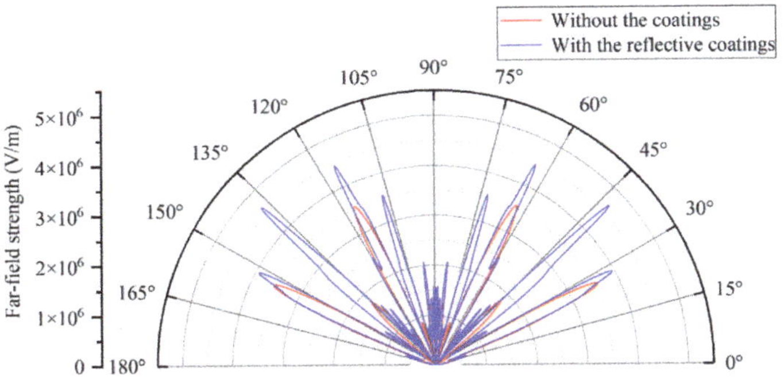

Figure 15. Far-field radiation pattern of multiple point sources.

We evaluated the reflective performance of the device under varying spacing between point sources, as shown in Figure 16. The results show that even in the lowest-performing configuration, the reflection efficiency remains close to 40%, and the power enhancement exceeds 60%. This indicates that the proposed structure exhibits stable reflective performance under different point source arrangements, effectively enhancing ultraviolet emission across a range of excitation conditions.

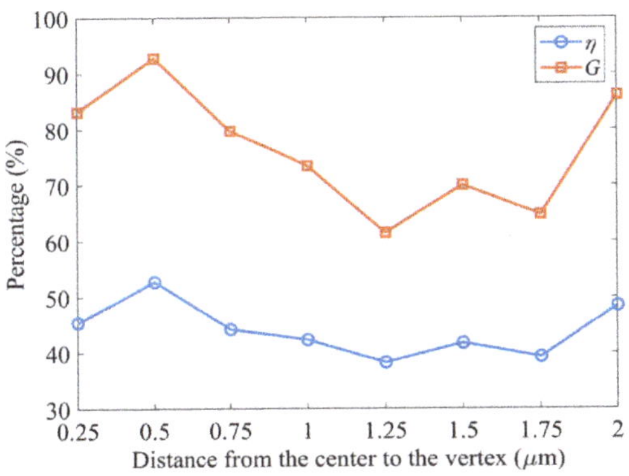

Figure 16. Effects of point source spacing on reflective performance.

We fabricate the V-shaped grooves and deposit the multilayer reflective coatings. From Figure 17, one can see that the proposed device is successfully fabricated with highly uniform and conformal coatings formed. In practical implementations, arrays containing hundreds or thousands of such grooves will be fabricated on a chip, providing cumulative UV output sufficient for applications such as disinfection and sensing. However, due to the lack of some necessary equipment, ultraviolet microplasma excitation cannot be conducted at this stage. When operated in microplasma mode, the coatings may have performance degradation due to possible heating, sputtering, transformation, and contamination.

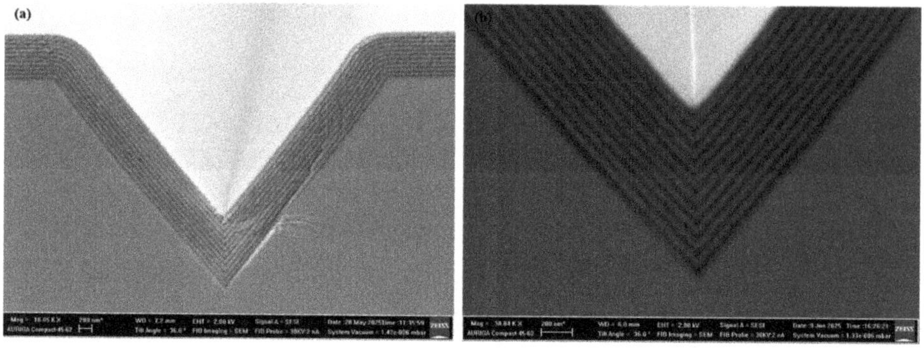

Figure 17. Deposition results of the reflective multilayer coatings. (**a**) Characterization of the V-shaped groove under scanning electron microscope. (**b**) Detailed view.

4. Conclusions

In summary, this paper presents the design of a V-shaped groove integrated with a reflective multilayer coating and demonstrates that groove width plays a critical role in determining the control capability of the optical field. At a groove width of 6.5 μm, the structure achieves optimal reflective performance, with a reflection efficiency of up to 47.82% and a power enhancement of 91.66%, corresponding to improvements of approximately 2.5-fold and 4.2-fold, respectively, compared to non-optimized groove widths. In addition, the offset of the light source significantly affects reflective performance, with offsets along the x-axis causing particularly large fluctuations. Therefore, in practical applications, it is essential to prioritize the precise alignment of the light source along the x-axis to maintain structural symmetry and consistency of reflection paths. Meanwhile, moderate adjustments of the y-axis position can further optimize the far-field radiation direction and enhance the control over energy, thereby achieving efficient and stable ultraviolet emission enhancement.

Author Contributions: Conceptualization, L.Z.; validation, H.Z.; formal analysis, H.Z.; investigation, H.Z. and L.Z.; resources, L.Z. and Z.H.; data curation, H.Z.; writing—original draft preparation, H.Z.; writing—review and editing, L.Z.; supervision, L.Z. and Z.H.; project administration, L.Z. All authors have read and agreed to the published version of the manuscript.

Funding: This research received no external funding.

Institutional Review Board Statement: Not applicable.

Informed Consent Statement: Not applicable.

Data Availability Statement: Data are contained within the article.

Acknowledgments: We acknowledge the support from the Advanced Integrated Optoelectronics Facility at Tianjin University.

Conflicts of Interest: The authors declare no conflicts of interest.

References

1. Lin, L.; Wang, Q. Microplasma: A new generation of technology for functional nanomaterial synthesis. *Plasma Chem. Plasma Process.* **2015**, *35*, 925–962. [CrossRef]
2. Zheng, L.; Kulkarni, P.; Diwakar, P. Spatial and temporal dynamics of a pulsed spark microplasma used for aerosol analysis. *Spectrochim. Acta Part B* **2018**, *144*, 55–62. [CrossRef]
3. Chiang, W.H.; Mariotti, D.; Sankaran, R.M.; Eden, J.G.; Ostrikov, K. Microplasmas for advanced materials and devices. *Adv. Mater.* **2020**, *32*, 1905508. [CrossRef] [PubMed]

4. Park, S.J.; Chen, J.; Wagner, C.J.; Ostrom, N.P.; Liu, C.; Eden, J.G. Microdischarge arrays: A new family of photonic devices. *IEEE J. Sel. Top. Quantum Electron.* **2002**, *8*, 387–394. [CrossRef]
5. Becker, K.H.; Schoenbach, K.H.; Eden, J.G. Microplasmas and applications. *J. Phys. D Appl. Phys.* **2006**, *39*, R55. [CrossRef]
6. Schoenbach, K.H.; Becker, K. 20 years of microplasma research: A status report. *Eur. Phys. J. D* **2016**, *70*, 29. [CrossRef]
7. Nijdam, S.; Desai, K.V.; Park, S.J.; Sun, P.P.; Sakai, O.; Lister, G.; Eden, J.G. Foundations of plasma photonics: Lamps, lasers, and electromagnetic devices. *Plasma Sources Sci. Technol.* **2023**, *31*, 123001. [CrossRef]
8. Thaury, C.; Quere, F.; Geindre, J.P.; Levy, A.; Ceccotti, T.; Monot, P.; Bougeard, M.; Réau, F.; d'Oliveira, P.; Audebert, P.; et al. Plasma mirrors for ultrahigh-intensity optics. *Nat. Phys.* **2007**, *3*, 424–429. [CrossRef]
9. Leblanc, A.; Denoeud, A.; Chopineau, L.; Mennerat, G.; Martin, P.; Quéré, F. Plasma holograms for ultrahigh-intensity optics. *Nat. Phys.* **2017**, *13*, 440–443. [CrossRef]
10. Kunze, K.; Miclea, M.; Franzke, J.; Niemax, K. The dielectric barrier discharge as a detector for gas chromatography. *Spectrochim. Acta Part B* **2003**, *58*, 1435–1443. [CrossRef]
11. Stoffels, E.; Flikweert, A.J.; Stoffels, W.W.; Kroesen, G. Plasma needle: A non-destructive atmospheric plasma source for fine surface treatment of (bio) materials. *Plasma Sources Sci. Technol.* **2002**, *11*, 383. [CrossRef]
12. Laroussi, M.; Lu, X.P. Room-temperature atmospheric pressure plasma plume for biomedical applications. *Appl. Phys. Lett.* **2005**, *87*, 113902. [CrossRef]
13. Shirai, H.; Kobayashi, T.; Hasegawa, Y. Synthesis of silicon nanocones using rf microplasma at atmospheric pressure. *Appl. Phys. Lett.* **2005**, *87*, 143112. [CrossRef]
14. Anderson, M. The ultraviolet offense: Germicidal UV lamps destroy vicious viruses. New tech might put them many more places without harming humans. *IEEE Spectr.* **2020**, *57*, 50–55. [CrossRef]
15. Ma, B.; Bright, K.; Ikner, L.; Ley, C.; Seyedi, S.; Gerba, C.P.; Sobsey, M.D.; Piper, P.; Linden, K.G. UV inactivation of common pathogens and surrogates under 222 nm irradiation from KrCl* excimer lamps. *Photochem. Photobiol.* **2023**, *99*, 975–982. [CrossRef] [PubMed]
16. García de Abajo, F.J.; Hernández, R.J.; Kaminer, I.; Meyerhans, A.; Rosell-Llompart, J.; Sanchez-Elsner, T. Back to normal: An old physics route to reduce SARS-CoV-2 transmission in indoor spaces. *ACS Nano* **2020**, *14*, 7704–7713. [CrossRef]
17. Frame, J.W.; Wheeler, D.J.; DeTemple, T.A.; Eden, J.G. Microdischarge devices fabricated in silicon. *Appl. Phys. Lett.* **1997**, *71*, 1165–1167. [CrossRef]
18. Park, S.J.; Chen, J.; Liu, C.; Eden, J.G. Arrays of microdischarge devices having 50–100 μm square pyramidal Si anodes and screen cathodes. *Electron. Lett.* **2001**, *37*, 171–172. [CrossRef]
19. Eden, J.G.; Park, S.J. Microcavity plasma devices and arrays: A new realm of plasma physics and photonic applications. *Plasma Phys. Control. Fusion* **2005**, *47*, B83–B92. [CrossRef]
20. Wu, C.; Hoskinson, A.R.; Hopwood, J. Stable linear plasma arrays at atmospheric pressure. *Plasma Sources Sci. Technol.* **2011**, *20*, 045022. [CrossRef]
21. Eden, J.G.; Park, S.J.; Herring, C.M.; Bulson, J.M. Microplasma light tiles: Thin sheet lamps for general illumination. *J. Phys. D Appl. Phys.* **2011**, *44*, 224011. [CrossRef]
22. Wang, Y.; Ni, J.H.; Zhong, S.; Zhang, X.; Liang, Z.; Liu, C.; Park, S.J.; Eden, J.G. Microplasma mode transition and corresponding propagation characteristics controlled by manipulating electric field strength in a microchannel-cavity hybrid structure device. *J. Phys. D Appl. Phys.* **2016**, *49*, 415206. [CrossRef]
23. Park, S.J.; Herring, C.M.; Mironov, A.E.; Cho, J.H.; Eden, J.G. 25 W of average power at 172 nm in the vacuum ultraviolet from flat, efficient lamps driven by interlaced arrays of microcavity plasmas. *APL Photonics* **2017**, *2*, 041302. [CrossRef]
24. Dzikowski, S.; Michaud, R.; Böttner, H.; Dussart, R.; Böke, M.; Schulz-von Der Gathen, V. Modular constructed metal-grid arrays—An alternative to silicon-based microplasma devices for catalytic applications. *Plasma Sources Sci. Technol.* **2020**, *29*, 035028. [CrossRef]
25. Sun, J.; Li, X.; Zhang, W.; Yi, K.; Shao, J. High-reflectivity mirrors by Al_2O_3, LaF_3 and AlF_3 for 193 nm application. *Opt. Laser Technol.* **2014**, *56*, 65–70. [CrossRef]
26. Narukage, N.; Kubo, M.; Ishikawa, R.; Ishikawa, S.N.; Katsukawa, Y.; Kobiki, T.; Giono, G.; Kano, R.; Bando, T.; Tsuneta, S.; et al. High-reflectivity coatings for a vacuum ultraviolet spectropolarimeter. *Sol. Phys.* **2017**, *292*, 40. [CrossRef]
27. Kim, J.; Mironov, A.; Park, S.; Kim, C.; Park, S.J.; Eden, J.G. Ultrasound harmonic generation and atomic layer deposition of multilayer, deep-UV mirrors and filters with microcavity plasma arrays. *Eur. Phys. J. D* **2023**, *77*, 73. [CrossRef]
28. Jia, X.; Kapraun, J.; Wang, J.; Qi, J.; Ji, Y.; Chang-Hasnain, C. Metasurface reflector enables room-temperature circularly polarized emission from VCSEL. *OPTICA* **2023**, *10*, 1093–1099. [CrossRef]
29. Yuan, L.; Fan, S. Temporal modulation brings metamaterials into new era. *Light Sci. Appl.* **2022**, *11*, 173. [CrossRef]
30. Oton, E.; Cigl, M.; Morawiak, P.; Mironov, S.; Bubnov, A.; Piecek, W. All-optical 3D blue phase photonic crystal switch with photosensitive dopants. *Sci. Rep.* **2024**, *14*, 9910. [CrossRef]
31. Yu, D.; Tan, H. *Engineering Optics*, 4th ed.; China Machine Press: Beijing, China, 2015; pp. 376–380. ISBN 978-7-111-51962-1.

32. Boidin, R.; Halenkovič, T.; Nazabal, V.; Beneš, L.; Němec, P. Pulsed laser deposited alumina thin films. *Ceram. Int.* **2016**, *42*, 1177–1182. [CrossRef]
33. Rodríguez-de Marcos, L.V.; Larruquert, J.I.; Méndez, J.A.; Aznárez, J.A. Self-consistent optical constants of SiO_2 and Ta_2O_5 films. *Opt. Mater. Express* **2016**, *6*, 3622–3637. [CrossRef]
34. Berenger, J.P. A perfectly matched layer for the absorption of electromagnetic waves. *J. Comput. Phys.* **1994**, *114*, 185–200. [CrossRef]
35. Xie, S.; Pedrow, P.D. A novel active needle probe in an atmospheric pressure corona-based cold plasma reactor with admixtures of helium and dry air. *IEEE Trans. Plasma Sci.* **2020**, *48*, 2418–2430. [CrossRef]

photonics MDPI

Article

Optical Characteristics of GaAs Spherical Quantum Dots Based on Single and Double Quartic Anharmonic Potentials: The Role of Structural Parameters

Najah Abdullah Alashqar [1,2], Walid Belhadj [3], Najla S. Al-Shameri [2,4], Hassen Dakhlaoui [2,4,*], Fatih Ungan [5,6] and Sake Wang [7]

1 Physics Department, College of Science, Qassim University, Buraydah 52571, Saudi Arabia; 2240700042@iau.edu.sa
2 Physics Department, College of Science of Dammam, Imam Abdulrahman Bin Faisal University, P.O. Box 1982, Dammam 31441, Saudi Arabia; nalmasudi@iau.edu.sa
3 Physics Department, Faculty of Applied Science, Umm AL-Qura University, P.O. Box 715, Makkah 24382, Saudi Arabia; wbbelhadj@uqu.edu.sa
4 Nanomaterials Technology Unit, Basic and Applied Scientific Research Center (BASRC), Physics Department, College of Science of Dammam, Imam Abdulrahman Bin Faisal University, P.O. Box 1982, Dammam 31441, Saudi Arabia
5 Department of Physics, Faculty of Science, Sivas Cumhuriyet University, 58140 Sivas, Türkiye; unganf@gmail.com
6 Nanophotonics Application and Research Center, Sivas Cumhuriyet University, 58140 Sivas, Türkiye
7 College of Science, Jinling Institute of Technology, 99 Hongjing Avenue, Nanjing 211169, China; isaacwang@jit.edu.cn
* Correspondence: hbaldkhlaen@iau.edu.sa

Abstract

This is a numerical investigation of optical and electronic characteristics of GaAs spherical quantum dots based on single and double quartic potentials and presenting a hydrogenic impurity at their center. The radial Schrödinger equation was solved using the finite difference method (FDM) to obtain the energy levels and the wavefunctions. These physical quantities were then used to compute the dipole matrix elements, the total optical absorption coefficient (TOAC), and the binding energies. The impact of the structural parameters in the confining potentials on the red and blue shifts of the TOAC is discussed in the presence and absence of hydrogenic impurity. Our results indicate that the structural parameter k in both potentials plays a crucial role in tuning the TOAC. In the case of single quartic potential, increasing k produces a blue shift; however, its augmentation in the case of double quartic potential displays a blue shift at first, and then a red shift. Furthermore, the augmentation of the parameter k can control the binding energies of the two lowest states, $(1s)$ and $(1p)$. In fact, enlarging this parameter reduces the binding energies and converges them to constant values. In general, the modification of the potential's parameters, which can engender two shapes of confining potentials (single quartic and double quartic), enables the experimenters to control the desired energy levels and consequently to adjust and select the suitable TOAC between the two lowest energy states (ground $(1s)$ and first excited $(1p)$).

Keywords: optical absorption coefficient; binding energy; GaAs spherical quantum dot; Schrödinger equation; hydrogenic impurity

1. Introduction

It is well known that the harmonic oscillator potential (HOP) with different dimensions is considered the highest significant model that is largely studied in quantum mechanics problems [1–9]. Its importance is due to its excellent results in describing molecular vibrations between molecules around their equilibrium positions. Furthermore, the quantum systems under HOP generally present exact analytical solutions in one, two, and three dimensions. Recent studies have demonstrated that in many cases the motion is not a simple harmonic oscillation, and adding anharmonic potential to the harmonic one improved the explanation of the molecular vibrations [6–11]. The general form of anharmonic oscillator potential (AOP) as presented in different studies contains higher order terms such as the single and double quartic, the sextic, and the octic. The incorporation of these terms within the HOP improves the performance of the potential, especially when it describes larger displacement between molecules where simple oscillation is no longer efficient.

In recent years, many research works have concentrated on the electronic and optical properties of quantum systems based on anharmonic potentials in molecular and solid-state physics. Some of these works explored numerically [12–18], and others were analytical [19–21]. For instance, diatomic molecules are considered to be oscillating systems. In these systems, the nuclei motions usually present non-periodic oscillations, and they are often described by combining harmonic and anharmonic potentials. In this regard, when investigating the diatomic molecules' motion in quantum mechanics, the anharmonic potential illustrating the electrostatic interaction between atoms must involve higher-order terms such as sextic, quartic, and even decatic. Additionally, anharmonic potentials with higher anharmonicity have demonstrated a crucial role in studying quantum tunneling problems and molecular spectra [21–28]. Furthermore, recent experimental results have proposed that the confinement in spherical QDs can be better described by an anharmonic potential with a finite depth rather than a harmonic one [25,26]. The importance of anharmonic potential is not limited to solid state physics, but is also found in nuclear physics. For instance, Hamad et al. studied the critical fluctuations in the ground and excited states using a model based on anharmonic potential [29]. They proved that the quartic, sextic, and octic anharmonic oscillators can be used to describe yrast and non-yrast energy sequences in many even–even atoms such as ^{100}Mo, ^{146}Ba, and ^{226}Ra. The impact of variable electron mass on the optical properties in anharmonic oscillators was studied by Sakiroglu et al. [10]. They demonstrated that the energy levels and their separations can be modified due to the position dependence of the effective mass. Yucel et al. investigated the influence of an applied laser field on the optical properties in anharmonic oscillators with higher-order anharmonicity [9]. They also discussed the impact of the structural parameters such as the different sizes and coupling between the wells on the red and blue shift of the optical absorption. In our previous works [30,31], we studied the binding energies and optical absorption in spherical quantum dots based on power-exponential and Woods–Saxon potentials. Those potentials are largely employed to describe anharmonic interactions, especially in diatomic molecules. However, yrast and non-yrast energy sequences in even–even atoms such as ^{100}Mo, ^{146}Ba, and ^{226}Ra are better modelized by anharmonic potentials with higher orders, such as the single and double quartic oscillators.

In this context, the binding energy, electronic and optical properties of spherical GaAs quantum dots with single and double quartic potentials and in the presence of on-center hydrogenic impurity are investigated in the present study by changing the structural parameters governing the geometric shape of the confining potential.

2. Theory

2.1. Single and Double Quartic Potential Profiles

In this section, we first aim to present the geometrical forms of the studied confining potentials, which are the single and double quartic potentials. Their profiles' dependence on the structural parameters enables us to predict the spatial expansion of the 1s and 1p wavefunctions, as well as the tendency of their associated energies. This is essential when the studied system is a perfect spherical quantum dot (QD) displaying a spherical symmetry. Given the presence of this symmetry, the electron wavefunctions and their associated densities are computed by simply solving the radial part of the Schrödinger equation (RSE). This latter depends mainly on the geometric form of the potential. Allowing that the energy interval between the sub-band energy levels denotes a key aspect in the optical absorption, their tunability and control is made mandatory by applying external fields, varying the QD dimensions, or altering the confining potential's parameters. Herein, we consider a single and double quartic potential derived from the general form of the anharmonic potential by varying the parameters p_1 and p_2 as well as the variable k. Its analytical expression is given by the following expression [10]:

$$V_{Quartic}(r) = V_c\left[p_1\left(\frac{r}{k}\right)^2 + p_2\left(\frac{r}{k}\right)^4\right] \quad (1)$$

V_c represents the conduction band discontinuity between AlGaAs and GaAs semiconductors. p_1, p_2, and k are considered to be parameters of simulation. In Figure 1, we display the single quartic anharmonic potential for different values of the parameter k. This potential was obtained by selecting p_1 and p_2 equal to 0.2 and 0.5, respectively. The parabolicity of the potential increases with k, especially for a larger radius $\left(r > 30\ \text{Å}\right)$, which in turn affects the distribution of the energy levels E_{nl} and their radial wavefunctions $R_{nl}(r)$ and the (TOAC) between the 1s and 1p states.

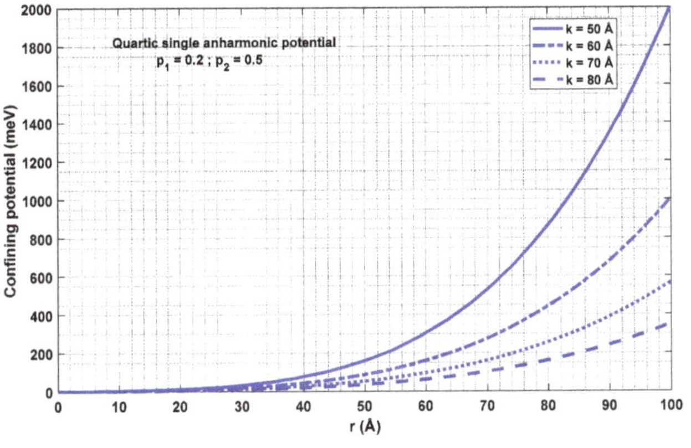

Figure 1. Single quartic single potential for different values of parameter k.

By selecting different values of the structural parameters than those used before, we can generate the double quartic confining potential. In the case where the parameters p_1 and p_2 are equal to -1 and 0.3, respectively, the anharmonic potential presents a parabolic barrier centered at the origin of the quantum dot separating two quantum wells (left and right). Since the potential is symmetric with respect to $(r = 0)$, in Figure 2 we represent only its form for positive radius r. By examining Figure 2, we can observe that the augmentation of k widens the parabolicity of the confining potential without affecting its minimum value, which is around -200 meV.

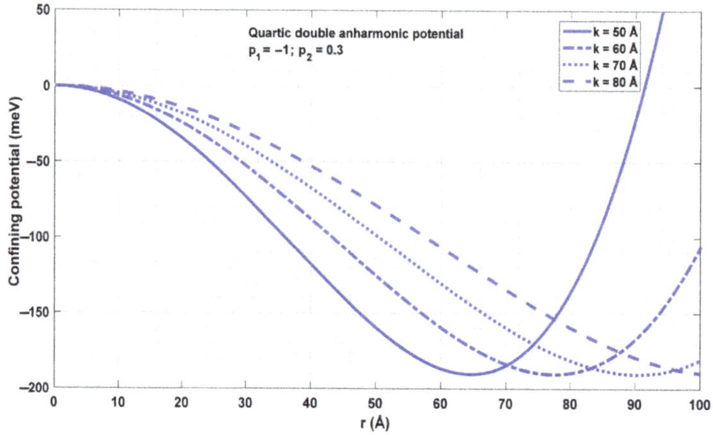

Figure 2. Double quartic potential for different values of parameter k.

2.2. Radial Schrödinger's Equations

Retaining the quartic potential term showed by Equation (1), the different wavefunctions $R_{n\ell}(r)$ and energy levels $E_{n\ell}$ of the spherical GaAs QD can be computed numerically by solving the following equation [30,31]:

$$\left[-\frac{\hbar^2}{2}\vec{\nabla}_r\left(\frac{1}{m^*(r)}\vec{\nabla}_r\right)+\frac{\ell(\ell+1)\hbar^2}{2m^*(r)r^2}-\frac{Ze^2}{\varepsilon r}+V_{Quartic}(r)\right]R_{n\ell}(r)=E_{n\ell}R_{n\ell}(r) \tag{2}$$

In the above equation, $\hbar$ represents the reduced Planck constant, $m^*(r)$ stands for the position-dependent mass of the electron, ε denotes the dielectric constant, Z represents the impurity charge, and ℓ stands for the angular quantum number.

The first term of Equation (2) denotes the kinetic energy of the free electron, the second term denotes the centrifugal potential, and the quartic potential energy is described by the last term of Equation (2). The attractive potential between the free electron and hydrogenic impurity is denoted by the term $\left(-\frac{Z\,e^2}{\varepsilon\,r}\right)$. The numerical procedure consists of discretizing Equation (2) using the (FDM) method and transforming it to a linear problem of the type $(\xi u = \alpha u)$. α and u represent $E_{n\ell}$ and $R_{n\ell}(r)$, respectively. On the other hand, ξ is a tridiagonal matrix [30,31].

$$\xi_{ij}=\begin{cases}\frac{\hbar^2}{m^*(\Delta r)^2}+\frac{\ell(\ell+1)}{m^*(r_j\cdot\Delta r)^2}+V_{Anharmonic}(j),\ if\ j=i\\ \frac{\hbar^2}{2m^*r_j(\Delta r)}-\frac{\hbar^2}{2m^*(\Delta r)^2},\ if\ j=i-1\\ -\frac{\hbar^2}{2m^*r(\Delta r)}-\frac{\hbar^2}{2m^*(\Delta r)^2},\ if\ j=i+1\\ 0\ ;otherwise\end{cases} \tag{3}$$

where $r_j = j\Delta r$, $(j=1,\ldots,N)$, and $\Delta r = \frac{R}{N}$ denotes the step of discretization. The dimension of matrix ξ is $(N\times N)$. In all our simulations, we have taken $N = 1200$.

In agreement with Fermi's golden rule, the TOAC between two states (initial and final) can be derived using the next expression [30–32]:

$$\alpha(\hbar\omega)=\beta\hbar\omega\left|M_{if}\right|^2\delta\left(E_f-E_i-\hbar\omega\right) \tag{4}$$

with

$$\beta = \frac{16\pi^2 \gamma_{FS} N_{if}}{n_r V_{con}} \quad (5)$$

In our study, the delta function is approximated by the Lorentzian function [32]:

$$\delta\left(E_f - E_i - \hbar\omega\right) = \frac{\hbar\Gamma}{\pi\left[\left(E_f - E_i - \hbar\omega\right)^2 + (\hbar\Gamma)^2\right]} \quad (6)$$

Through the next sections of the paper, the final state $(f = 2)$ is the *1p* state and the initial state $(i = 1)$ is *1s*; consequently, the square value of the dipole matrix element $\left|M_{if}\right|^2$ in Equation (4) is represented by $|M_{12}|^2$, and given by the following expression [32]:

$$|M_{12}|^2 = \frac{1}{3}\left|\int_0^\infty R_f(r) r^3 R_i(r) dr\right|^2 \quad (7)$$

where $R_f(r)$ and $R_i(r)$ represent the radial wavefunction of the final (1p) and initial (1s) states, respectively. The factor $\frac{1}{3}$ comes from the integration of the spherical harmonics.

The atomic units were used in the present simulation [33–36]; therefore, the Bohr radius and Rydberg energy are equal to $1a_B \cong 100$ Å and $1R_y \cong 5.6$ meV, respectively. The other parameters are $V_c = 0.228$ eV; $R = 200$ Å; $\varepsilon = 13.11\varepsilon_0$; $m^* = 0.067m_0$; and $\hbar\Gamma = 3$ meV.

3. Results and Discussion

3.1. Single Quartic Anharmonic Oscillator

To highlight the parabolicity impact on the energy levels and to give further ideas on the energy levels' tendency when we increment the parameter k, in Figure 3, we plot the lower four levels $(E_{1s}, E_{2s}, E_{1p}$, and $E_{2p})$ depending on k. It is clear that all the energy levels decrease upon augmenting k, which is legitimate since the potential widens its width, giving greater confinement and consequently lowering all the confined energy states (see Figure 1). In addition, we observed that the energy levels in the presence of hydrogenic impurity were less than those without its presence. The small energetic shift introduced by the presence of the hydrogenic impurity is due to attraction between impurity and free electrons.

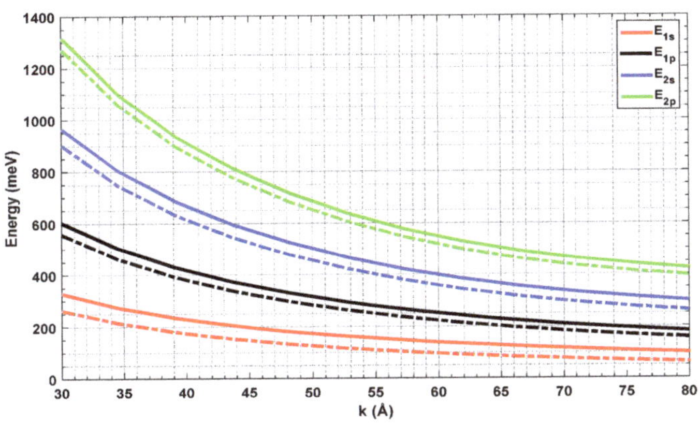

Figure 3. Energy level variation as a function of k (dashed lines (Z = 1); solid lines (Z = 0)).

Since the optical absorption coefficient α_{1s-1p} of the transition $(1s \rightarrow 1p)$ is at the core of the present study, in Figure 4, we plot its variation as a function of the incident photon energy for four values of k. It appears that the absorption peak moves to lower energies by incrementing k suffering; therefore, red-shift behavior is obtained. Furthermore, the absorption's amplitude rises from 3×10^4 cm^{-1} for $\left(k = 50\text{Å}\right)$ to 8×10^4 cm^{-1} for $\left(k = 80\text{Å}\right)$. The jump in amplitude and shift to the red of α_{1s-1p} is explained by the variation in $(E_{1p} - E_{1s})$ and the dipole matrix element (DME) given by the change in $|M_{12}|^2$. These physical quantities are displayed in Figure 5 for different values of the structural parameter k.

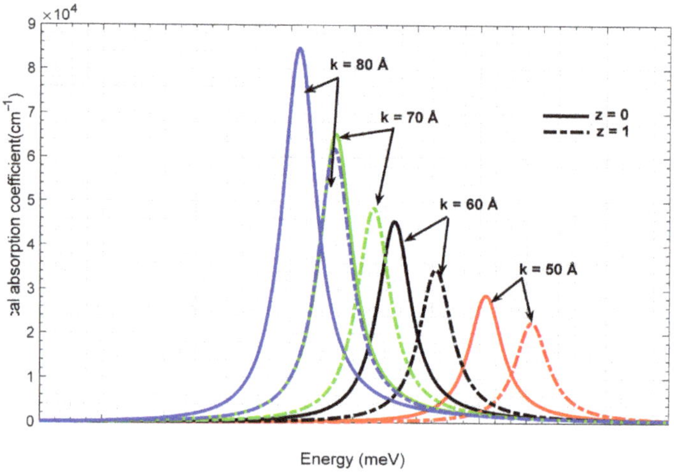

Figure 4. TOAC $\alpha_{1s-1p}(\omega)$ as a function of incident photon energy for different values of k.

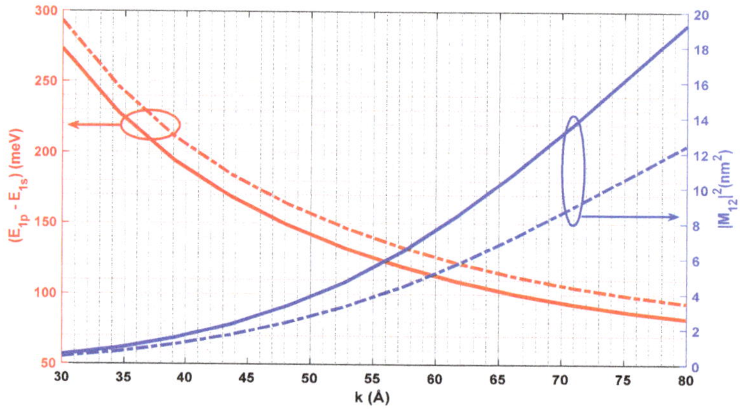

Figure 5. Dipole matrix element $|M_{12}|^2$ and energy variation for different values of k (dashed lines (Z = 1); solid lines (Z = 0)).

As shown in Figure 5, the energy separation $\Delta E = \left(E_{1p} - E_{1s}\right)$ diminishes with the parameter k, which confirms the red shift of the absorption's peak observed in Figure 4. In the meantime, we observe that contrary to the energy separation variation, the dipole matrix element $|M_{12}|^2$ augments upon increasing k, which confirms the increase in the absorption amplitude stated in Figure 4. The increase observed in $|M_{12}|^2$ is a result of the augmentation in the overlap between the radius r and the two wavefunctions $R_{1s}(r)$ and $R_{1p}(r)$. Figure 6 shows the spatial distribution of $R_{1s}(r)$, $R_{2s}(r)$, and $R_{1p}(r)$ for different values of k. By examining Figure 5, we can observe that the spread of the wavefunctions along the radial coordinate r augments with the parameter k. This augmentation in the spread leads to a greater overlap between $R_{1s}(r)$ and $R_{1p}(r)$, which confirms the increase of $|M_{12}|^2$ as presented in Figure 5. Note that the dependency on the hydrogenic impurity presence is clearly noticeable, especially for 1s and 2s states at the origin of the QD. In fact, with the presence of the impurity, $R_{1s}(r)$ and $R_{2s}(r)$ are larger than their values in the absence of the impurity, especially at the center of the QD ($r = 0$). In fact, the presence of the impurity reinforces the presence of lower states and consequently increases their wavefunctions and densities due to the attractive force between the impurity and free electrons. Note that the impact of the hydrogenic impurity in the center of the QD, especially on the shift of the TOAC and its amplitude due to the huge electrostatic attraction between the impurity and the carriers, was confirmed in previous works [36,37]. In fact, the hydrogenic impurity plays a crucial role in the spread of the wavefunctions and their overlap, which in turn determines the value of the dipole matrix elements and energy separations.

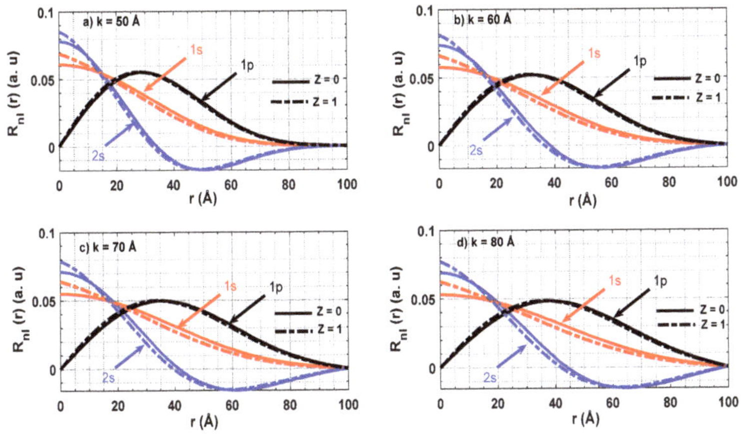

Figure 6. $R_{nl}(r)$ variation in the three lowest states (1*s*, 2*s*, and 1*p*) for four different values of *k*.

In Figure 7, we display the variation in the binding energies E_{1s} and E_{1p} as a function of the parameter *k*. We noted that these energies diminished upon increasing *k*. This diminution reflects the huge attraction generated by the impurity located at the QD center. The ground state (1s) has greater values than the 1p state for all values of *k*. This can be interpreted by the fact that the 1s state is close to the impurity position, which amplifies the electrostatic attraction between them, leading to large binding energy.

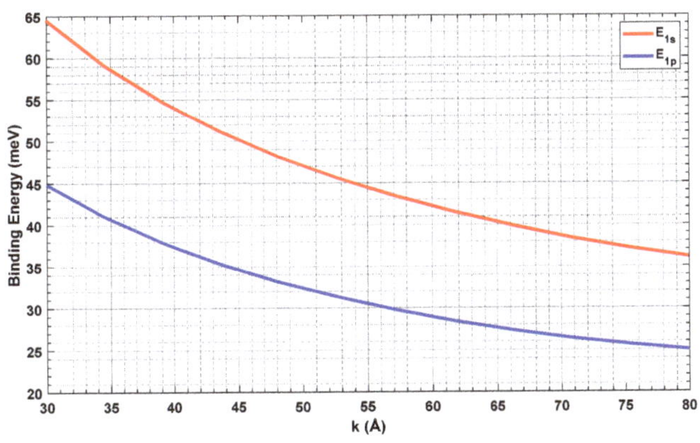

Figure 7. Binding energy variation as a function of *k*.

3.2. Double Quartic Anharmonic Oscillator

The geometrical changes observed in the double quartic potential will certainly affect the distribution of the energy levels. To see the effect of geometrical changes in potential on the energy levels' distribution, in Figure 8, we plot the lowest four levels as a function of k. By examining Figure 8, we can observe that the energy levels diminish rapidly for all $\left(k < 50\text{Å}\right)$ and then increase slowly. Furthermore, we observe that the energy of lower states $\left(E_{1s} \text{ and } E_{1p}\right)$ is less affected by the geometrical changes in the confining potential.

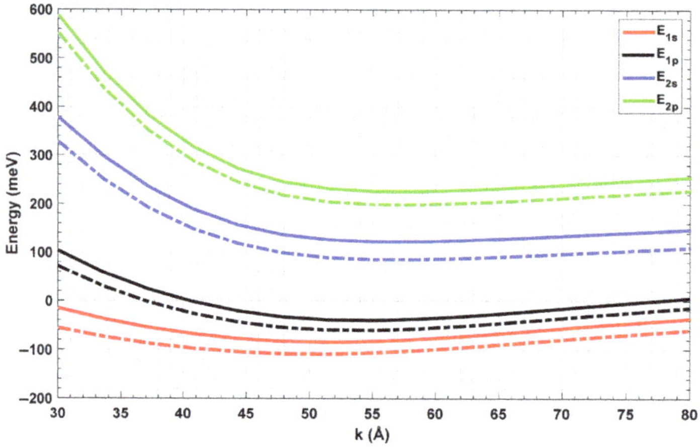

Figure 8. Energy level variation as a function of k (dashed lines (Z = 1); solid lines (Z = 0)).

The variation in $\alpha_{1s-1p}(\omega)$ as a function of incident photon energy for different values of k is given by Figure 9. We observe that the absorption's peak moves to lower energies when k is increased from 50Å to 70Å, displaying red shift behavior. However, when k surpasses 70Å to reach 80Å, the absorption's peak moves towards higher energies and a blue shift is observed. This result shows that the double quartic confining potential produces the two shifts (red and blue) of the TOAC by increasing the parameter k. However, the augmentation of this parameter in the case of single quartic anharmonic potential produces only a red shift of the TOAC. Furthermore, we can see clearly that the amplitude of the absorption increases at first and then diminishes. The shifts observed in the amplitudes and the peaks' positions are related to the energy separation $\left(E_{1p} - E_{1s}\right)$ and the dipole matrix element $|M_{12}|^2$. In Figure 10, we plot the variation in $\left(E_{1p} - E_{1s}\right)$ and $|M_{12}|^2$ as a function of the parameter k. This figure clearly shows that $|M_{12}|^2$ increases at first for all $\left(k < 70\text{Å}\right)$ and then decreases slowly. The change in $|M_{12}|^2$ confirms the amplitude's variation of $\alpha_{1s-1p}(\omega)$, as observed in Figure 9. Furthermore, the energy variation $\left(E_{1p} - E_{1s}\right)$ is displayed as a function of k. It diminishes at first for $\left(k < 70\text{Å}\right)$ and then rises gradually, which confirms the red and blue shifts of the TOAC detected in Figure 9. Note that for all values of k, the $|M_{12}|^2$ values in the presence of hydrogenic impurity are lower than those in the absence of it. This can be explained as follows: in the presence of hydrogenic impurity, the electron is located nearer the quantum dot center, so its wavefunction is reduced for a larger radius, and consequently the overlap between $R_{1s}(r)$ and $R_{1p}(r)$ is smaller than their overlap in the absence of the on-center impurity.

Finally, we show in Figure 11 the variation in the binding energies of the two lowest states $\left(E_{1s} \text{ and } E_{1p}\right)$ as a function of k. The binding energies decreases at first until a certain minimum obtained for $k = 55$ Å, and then remain constant. This is due to the insensitivity of the energy levels at higher values of k. In fact, when k is larger than 55 Å, the change in the bottom of the confining potential is reduced and the energy levels are not further affected.

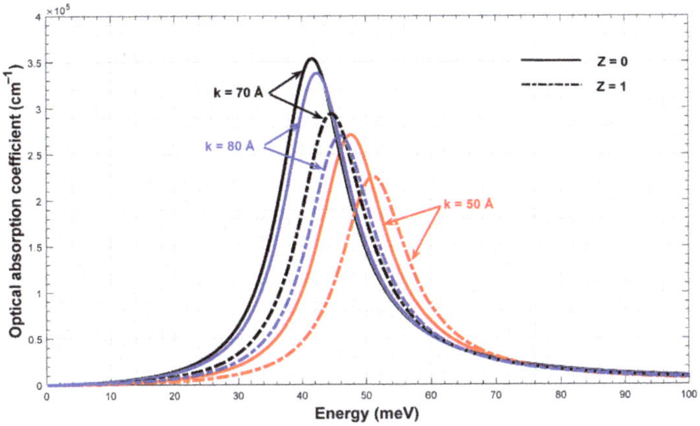

Figure 9. TOAC $\alpha_{1s-1p}(\omega)$ as function of incident photon energy for different values of k.

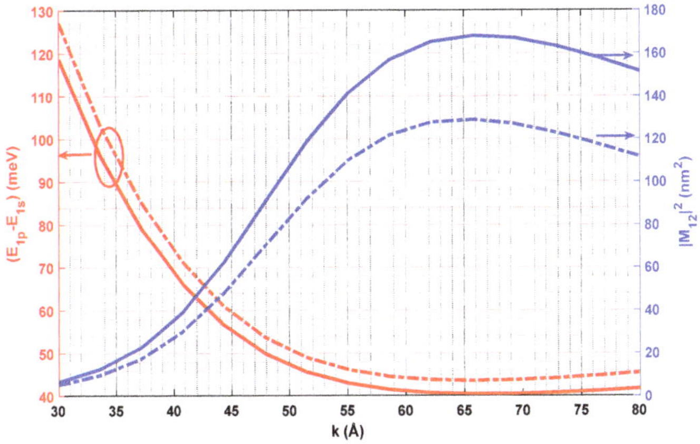

Figure 10. Diploe matrix element $|M_{12}|^2$ and energy variation $(E_{1p} - E_{1s})$ for different values of k (dashed lines (Z = 1); solid lines (Z = 0)).

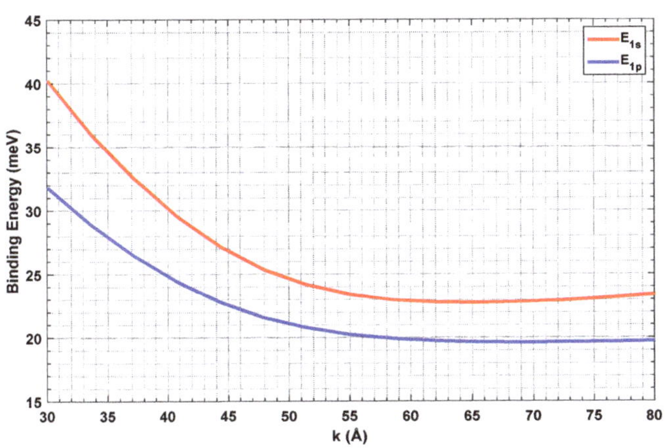

Figure 11. Binding energy variation as a function of k.

4. Conclusions

As is well known, the vibrational interactions between atoms and molecules are better explained by using anharmonic potentials, which are more consistently realistic than harmonic ones, to describe these interactions. The efficiency of anharmonic (single and double quartic) potentials is obtained especially when studying diatomic molecules such as AHO, where the energy levels are not equally spaced. Therefore, because of these potentials' importance, we have studied the electronic and optical properties of on-center hydrogenic spherical GaAs quantum dots based on single and double quartic potentials. The wavefunctions obtained by solving Schrödinger's Equation were used to compute the dipole matrix elements and the energy separations, as well as the TOAC and the binding energies. Our results indicate that the structural parameter k in both potentials plays a crucial role in tuning the TOAC. In the case of single quartic potential, increasing k produces a blue shift; however, its augmentation in the case of double quartic potential displays a blue shift at first, and then a red shift. The binding energy dependence on parameter k was also commented on in the presence of each potential. The present study of the TOAC in spherical QDs with the presence of hydrogenic impurity can be useful for experimental applications in new devices based on anharmonic potentials.

Author Contributions: N.A.A.: Conceptualization, Formal analysis, Investigation, Software, Writing—original draft. W.B.: Conceptualization, Formal analysis, Investigation, Software, Writing—original draft. N.S.A.-S.: Formal analysis, Investigation, Software. H.D.: Conceptualization, Formal analysis, Investigation, Software, Writing—original draft. F.U.: Conceptualization, Formal analysis, Investigation, Software. S.W.: Conceptualization, Formal analysis, Investigation, Software, Writing—original draft. All authors have read and agreed to the published version of the manuscript.

Funding: This research work was funded by Umm Al-Qura University, Saudi Arabia, under grant number 25UQU4331235GSSR04.

Institutional Review Board Statement: Not applicable.

Informed Consent Statement: Not applicable.

Data Availability Statement: This work is new, and all the results were computed from the equations. The data that support the findings of this study are available from the corresponding author upon reasonable request.

Acknowledgments: The authors extend their appreciation to Umm Al-Qura University, Saudi Arabia for funding this research work through grant number: 25UQU4331235GSSR04.

Conflicts of Interest: The authors declare no conflicts of interest.

References

1. Villegas-Martínez, B.M.; Moya-Cessa, H.M.; Soto-Eguibar, F. Exact and approximated solutions for the harmonic and anharmonic repulsive oscillators: Matrix method. *Eur. Phys. J. D* **2020**, *74*, 137. [CrossRef]
2. Panek, T.; Hoeske, A.A.; Jacob, C.R. On the choice of coordinates in anharmonic theoretical vibrational spectroscopy: Harmonic vs. anharmonic coupling in vibrational configuration interaction. *J. Chem. Phys.* **2019**, *150*, 054107. [CrossRef] [PubMed]
3. Sharma, A.; Sastri, O.S.K.S. Numerical simulation of quantum anharmonic oscillator, embedded within an infinite square well potential, by matrix methods using Gnumeric spreadsheet. *Eur. J. Phys.* **2020**, *41*, 055402. [CrossRef]
4. Landi, G.T.; de Oliveira, M.J. Fourier's law from a chain of coupled anharmonic oscillators under energy-conserving noise. *Phys. Rev. E* **2013**, *87*, 052126. [CrossRef]
5. Fletcher, N.H. Harmonic? Anharmonic? Inharmonic? *Am. J. Phys.* **2002**, *70*, 1205. [CrossRef]
6. Adelakun, A.O.; Dele, A.D. Solution of quantum anharmonic oscillator with quartic perturbation. *Adv. Phys. Theor. Appl.* **2014**, *27*, 38–43.
7. Sous, A.J. Solution for the eigenenergies of sextic anharmonic oscillator potential $V(x) = A_6x^6 + A_4x^4 + A_2x^2$. *Mod. Phys. Lett. A* **2006**, *21*, 1675–1682. [CrossRef]

8. Pathak, A. Generalized quantum anharmonic oscillator using an operator ordering approach. *J. Phys. A Math. Gen.* **2000**, *33*, 5607–5613. [CrossRef]
9. Yücel, M.B.; Kasapoglu, E.; Duque, C.A. Effects of Intense Laser Field on Electronic and Optical Properties of Harmonic and Variable Degree Anharmonic Oscillators. *Nanomaterials* **2022**, *12*, 1620. [CrossRef]
10. Sakiroglu, S.; Yücel, M.B.; Kasapoglu, E. The effects of the variable mass on the electronic and nonlinear optical properties of octic anharmonic oscillators. *Eur. Phys. J. Plus* **2023**, *138*, 946. [CrossRef]
11. Al Sdran, N.; Maiz, F. Airy function approach and Numerov method to study the anharmonic oscillator potentials $V(x) = Ax^{2\alpha} + Bx^2$. *AIP Adv.* **2016**, *6*, 065323. [CrossRef]
12. Maiz, F.; Alqahtani, M.M.; Al Sdran, N.; Ghnaim, I. Sextic and decatic anharmonic oscillator potentials: Polynomial solutions. *Phys. B Condens. Matter* **2018**, *530*, 101–105. [CrossRef]
13. Gaudreau, P.; Slevinsky, R.M.; Safouhi, H. An asymptotic expansion for energy eigenvalues of anharmonic oscillators. *Ann. Phys.* **2013**, *337*, 261–277. [CrossRef]
14. Shah, B.P. Bound state eigenfunctions of an anharmonic oscillator in one dimension: A Numerov method approach. *Int. J. Adv. Appl. Math. Mech.* **2014**, *2*, 102–109.
15. Budaca, R.; Buganu, P.; Chabab, M.; Lahbas, A.; Oulne, M. Extended study on a quasi-exact solution of the Bohr Hamiltonian. *Ann. Phys* **2016**, *375*, 65–90. [CrossRef]
16. Ciftci, H.; Hall, R.L.; Saad, N. Construction of exact solutions to eigenvalue problems by the asymptotic iteration method. *J. Phys. A. Math. Gen.* **2005**, *38*, 1147–1155. [CrossRef]
17. Maiz, F.; Alqahtani, M.M.; Ghnaim, I. Sextic and decatic anharmonic oscillator potentials including odd power terms: Polynomial solutions. *AIP Con. Proc.* **2018**, *1976*, 020034.
18. Gaudreau, P.J.; Slevinsky, R.M.; Safouhi, H. Computing energy eigenvalues of anharmonic oscillators using the double exponential Sinc collocation method. *Ann. Phys.* **2015**, *360*, 520–538. [CrossRef]
19. Maiz, F.; AlFaify, S. Quantum anharmonic oscillator: The airy function approach. *Phys. B Cond. Matt.* **2014**, *441*, 17–20. [CrossRef]
20. Sous, A.J. The Asymptotic Iteration Method for the Eigenenergies of the a Novel Hyperbolic Single Wave Potential. *J. Appl. Math. Phys.* **2015**, *3*, 1406–1411. [CrossRef]
21. de Souza Dutra, A.; de Castro, A.S.; Boschi-Filho, H. Approximate analytic expression for the eigenenergies of the anharmonic oscillator $V(x) = Ax^6 + Bx^2$. *Phys. Rev. A* **1995**, *51*, 3480–3484. [CrossRef] [PubMed]
22. Fernandez, F.M.; Guardiola, R. Accurate eigenvalues and eigenfunctions for quantum-mechanical anharmonic oscillators. *J. Phys. A Math. Gen.* **1993**, *26*, 7169. [CrossRef]
23. Ivanov, I.A. Sextic and octic anharmonic oscillators: Connection between strong-coupling and weak-coupling expansions. *J. Phys. A Math. Gen.* **1998**, *31*, 5697. [CrossRef]
24. Jafarpour, M.; Afshar, D. Calculation of energy eigenvalues for the quantum anharmonic oscillator with a polynomial potential. *J. Phys. A Math. Gen.* **2002**, *35*, 87. [CrossRef]
25. Sharma, L.K.; Fiase, J.O. Non-perturbative energy expressions for the generalized anharmonic oscillator. *Eur. J. Phys.* **2000**, *21*, 167. [CrossRef]
26. Speliotopoulos, A.D. The general structure of eigenvalues in nonlinear oscillators. *J. Phys. A Math. Gen.* **2000**, *33*, 3809. [CrossRef]
27. Ikhdair, S.M.; Sever, R. Am alternative simple solution of the sextic anharmonic oscillator and perturbed coulomb problems. *Int. J. Mod. Phys. C* **2007**, *18*, 1571–1581. [CrossRef]
28. Barakat, T. The asymptotic iteration method for the eigenenergies of the anharmonic oscillator potential $V(x) = Ax^{2\alpha} + Bx^2$. *Phys. Lett. A* **2005**, *344*, 411–417. [CrossRef]
29. Hammada, M.M.; Yahiaa, M.M.; Motawehb, H.A.; Domac, S.B. Critical potentials and fluctuations phenomena with quartic, sextic, and octic anharmonic oscillator potentials. *Nucl. Phys. A* **2020**, *1004*, 122036. [CrossRef]
30. Alauwaji, R.M.; Dakhlaoui, H.; Algraphy, E.; Ungan, F.; Wong, B.M. Binding Energies and Optical Properties of Power-Exponential and Modified Gaussian Quantum Dots. *Molecules* **2024**, *29*, 3052. [CrossRef]
31. Dakhlaoui, H.; Belhadj, W.; Elabidi, H.; Ungan, F.; Wong, B.M. GaAs Quantum Dot Confined with a Woods–Saxon Potential: Role of Structural Parameters on Binding Energy and Optical Absorption. *Inorganics* **2023**, *11*, 401. [CrossRef]
32. Tas, H.; Sahin, M. The inter-sublevel optical properties of a spherical quantum dot-quantum well with and without a donor impurity. *J. Appl. Phys.* **2012**, *112*, 053717. [CrossRef]
33. Altuntas, I.; Dakhlaoui, H.; Mora-Ramos, M.E.; Ungan, F. Combined effects of electric, magnetic, and intense terahertz laser fields on the nonlinear optical properties in GaAs/GaAlAs quantum well with exponentially confinement potential. *Eur. Phys. J. Plus* **2023**, *136*, 1174. [CrossRef]
34. Dakhlaoui, H.; Altuntas, I.; Mora-Ramos, M.E.; Ungan, F. Numerical simulation of linear and nonlinear optical properties in heterostructure based on triple Gaussian quantum wells: Effects of applied external fields and structural parameters. *Eur. Phys. J. Plus* **2021**, *136*, 894. [CrossRef]

35. Dakhlaoui, H. Influence of doping layer concentration on the electronic transitions in symmetric $Al_xGa_{(1-x)}N/GaN$ double quantum wells. *Opt.—Int. J. Light Electron Opt.* **2013**, *124*, 3726–3729. [CrossRef]
36. Ed-Dahmouny, A.; Arraoui, R.; Jaouane, M.; Fakkahi, A.; Sali, A.; Es-Sbai, N.; El-Bakkari, K.; Zeiri, N.; Duque, C.A. The influence of the electric and magnetic fields on donor impurity electronic states and optical absorption coefficients in a core/shell $GaAs/Al_{0.33}Ga_{0.67}As$ ellipsoidal quantum dot. *Eur. Phys. J. Plus* **2023**, *138*, 774. [CrossRef]
37. Ed-Dahmouny, A.; Sali, A.; Es-Sbai, N.; Arraoui, R.; Duque, C.A. The impact of hydrostatic pressure and temperature on the binding energy, linear, third-order nonlinear, and total optical absorption coefficients and refractive index changes of a hydrogenic donor impurity confined in $GaAs/Al_xGa_{1-x}As$ double quantum dots. *Eur. Phys. J. Plus* **2022**, *137*, 784. [CrossRef]

hv photonics

MDPI

Article

Global–Local Cooperative Optimization in Photonic Inverse Design Algorithms

Mingzhe Li [1], Tong Wang [1], Yi Zhang [1], Yulin Shen [1], Jie Yang [1], Ke Zhang [1], Dehui Pan [1] and Ming Xin [1,2,*]

[1] School of Electrical and Information Engineering, Tianjin University, Tianjin 300072, China; lmz804472764@tju.edu.cn (M.L.); wttesp@tju.edu.cn (T.W.); zy_1@tju.edu.cn (Y.Z.); shenyl@tju.edu.cn (Y.S.); yangjiewg@tju.edu.cn (J.Y.); zhangke2672@tju.edu.cn (K.Z.); panda@tju.edu.cn (D.P.)

[2] Tianjin Key Laboratory of Brain-Inspired Intelligence Technology, Tianjin 300072, China

* Correspondence: xinm@tju.edu.cn

Abstract

We developed the Global–Local Integrated Topology inverse design algorithm (denoted as the GLINT algorithm), which employs a trajectory-based optimization strategy with waveguide–substrate material-flipping structural modifications, enabling the direct optimization of discrete waveguide–substrate binary structures. Compared to the conventional Direct Binary Search (DBS), the GLINT algorithm not only significantly enhances computational efficiency through its global search–local refinement framework but also achieves a superior 20 nm × 20 nm optimization resolution while maintaining its optimization speed—substantially advancing the design capability. Utilizing this algorithm, we designed and experimentally demonstrated a 3.5 μm × 3.5 μm dual-port wavelength division multiplexer (WDM), achieving a minimum crosstalk of −11.3 dB and a 2 μm × 2 μm 90-degree bending waveguide exhibiting a 0.31–0.52 dB insertion loss over the 1528–1600 nm wavelength range, both fabricated on silicon-on-insulator (SOI) wafers. Additionally, a 4.5 μm × 4.5 μm three-port WDM structure was also designed and simulated, demonstrating crosstalk as low as −36.5 dB.

Keywords: integrated photonic devices; inverse design; subwavelength structure

1. Introduction

Integrated photonic devices are crucial for advancing modern optical technologies, facilitating breakthroughs in high-speed communication [1], quantum information processing [2], and biosensing [3]. Traditional design strategies often focus on optimizing a limited set of geometric parameters, such as waveguide widths or cavity radii, which are typically confined to predefined regular shapes. Although this approach is straightforward, it significantly restricts design freedom by confining solutions to low-dimensional design spaces that struggle to meet the increasing demand for compact, high-performance devices capable of exploiting complex light–matter interactions. To address these limitations, inverse design methodologies have emerged as a transformative paradigm [4], utilizing computational optimization to explore high-dimensional design spaces and unlock photonic devices with unprecedented functionality, enabling innovations from the three-channel wavelength demultiplexer [5] to meta-dispersion microresonators [6] and non-reciprocal pulse routers [7]. These advancements highlight the paradigm's capability to surpass traditional design constraints, establishing it as an essential tool for next-generation integrated photonics [8–13].

Despite these achievements, current mainstream adjoint-based inverse design approaches utilizing density topology optimization (DTO) [14–16] face a critical challenge that substantially constrains the optimization efficiency. The inherent assumption of a spatially continuous refractive index distribution within this framework fundamentally conflicts with practical fabrication requirements. The mandatory binarization process for discretizing continuous refractive indices may completely negate the carefully optimized optical characteristics, thereby necessitating optimization reinitialization, which significantly compromises the design efficiency. While level-set optimization (LST) [17,18] algorithms are commonly combined with DTO [5,18,19] to mitigate this issue through the direct boundary optimization of binary structures, their standalone application suffers from critical limitations. Specifically, LST methods can only modify existing boundaries or internal cavities but cannot spontaneously generate new void structures within solid regions, thus requiring complementary optimization strategies for a practical implementation.

Alternative approaches such as direct binary search (DBS) [20–26] algorithms and rotatable DBS [27] variants circumvent the continuous refractive index assumption by discretizing the design space into rectangular or circular pixels. However, such pixel-based methods suffer from two critical drawbacks. First, these methods sacrifice substantial design freedom compared to DTO approaches, with typical pixel dimensions (~100 nm × 100 nm) being excessively large, severely constraining their optimization flexibility. Second, prohibitive computational demands stemming from exhaustive pixel-by-pixel searching in DBS algorithms render these methods impractical for large-scale or functionally complex photonic devices.

In this work, we introduce a Global–Local Integrated Topology inverse design algorithm, denoted throughout this work as the GLINT algorithm or the GLINT inverse design algorithm, that addresses these challenges through an iterative optimization framework. Our approach strategically combines a global search for a rapid performance improvement and local refinement for precision tuning. The global search phase rapidly identifies promising design regions where structural modifications yield significant performance gains, while the local refinement phase progressively refines geometries within these regions through meticulous parameter adjustments. When the local refinement stagnates, the algorithm automatically reactivates the global search to locate new optimization zones, cycling between these modes until target specifications are achieved. Crucially, our method not only eliminates the continuous refractive index assumption inherent in DTO algorithms but also, compared to the DBS algorithm, significantly enhances the design freedom and computational efficiency.

To validate our approach, we experimentally demonstrate its efficacy through the realization of two compact devices: a two-port wavelength division multiplexer (WDM) and a 90-degree bending waveguide. Additionally, we numerically design a three-port WDM, showcasing the algorithm's capability to handle increasingly complex design challenges. Remarkably, the 90-degree bending waveguide occupies a footprint of just 2 μm × 2 μm, while the two-port and three-port WDMs measure 12.25 μm^2 and 20.25 μm^2, respectively. These results highlight the algorithm's ability to achieve ultra-compact device geometries, significantly reducing spatial requirements for chip layouts and paving the way for large-scale, densely integrated photonic circuits.

2. The Principle of the GLINT Inverse Design Algorithm

The underlying principle of the GLINT inverse design algorithm can be illustrated through a WDM design example. In this example, all simulations employed a 3D finite-difference time-domain (3D-FDTD) methodology with a spatial resolution of 20 nm × 20 nm, uniform perfectly matched layer (PML) boundary conditions, and transverse-electric (TE) polarized excitation sources. The optimization domain is defined with dimensions of 3.5 μm × 3.5 μm × 220 nm. The substrate material is silicon dioxide (represented in white), and the waveguide material is silicon (represented in blue). To facilitate global search and local refinement operations, the optimization domain is subdivided into a uniform grid with a cell size of 20 nm × 20 nm. Notably, the grid size (20 nm × 20 nm in this work) defines the minimum feature size of the optimized structures. While this parameter is not fixed and can be scaled according to fabrication capabilities, all designs in this study adopt 20 nm × 20 nm features to rigorously validate the GLINT algorithm's performance under near-lithographic-resolution constraints. Figure 1a illustrates the initial structure, where the silicon (white) and silica (blue) regions are clearly defined. Two input signals at λ_1 = 1607.5 nm and λ_2 = 1528.8 nm enter from the left port, with the design objective of routing λ_1 to output port 1 and λ_2 to port 2. Notably, GLINT allows the free selection of initial structures—including fully waveguide- or substrate-filled configurations. However, better initial structures could reduce the optimization time. To assess the device's performance, the fundamental principle governing the Figure of Merit (*FOM*) formulation dictates that as the performance of the optimized structure asymptotically approaches the target performance, the *FOM* converges toward 0; conversely, the deviation from the target performance drives the *FOM* toward 1. For this work, we define the *FOM* as

$$FOM = 1 - \frac{P_1(\lambda_1)}{P_{in}(\lambda_1)} \frac{P_2(\lambda_2)}{P_{in}(\lambda_2)} \quad (1)$$

where $P_{in}(\lambda_i)$ is the input power at wavelength λ_i, and $P_i(\lambda_i)$ is the output power at λ_i from port i (i = 1,2). The optimization target aims to minimize this *FOM* (ideally to zero) through iterative structural modifications. After each iteration, the updated *FOM* value is denoted as FOM_0.

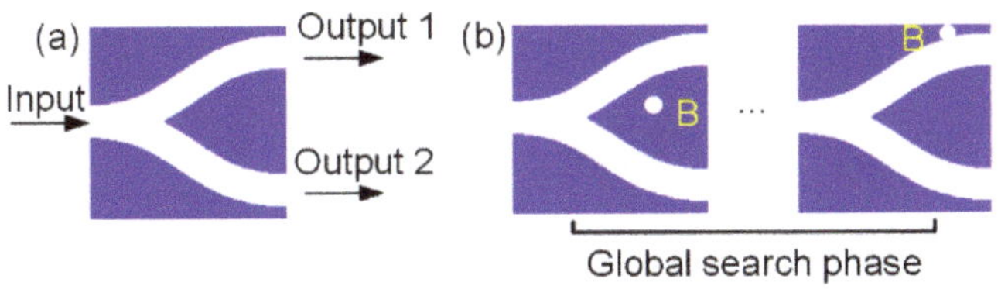

Figure 1. (**a**) The initial structure of the GLINT algorithm; (**b**) the global search phase; during this phase, the algorithm continuously searches for region **B** which satisfies condition (2).

As shown in Figure 1b, the algorithm initiates the global search by randomly selecting the circular region **B** within the design area (in this case, the radius of region **B** is 80 nm). The temporary material inversion (silicon ↔ silica) within **B** generates a trial structure, followed by the trial *FOM* calculation (FOM_t). After restoring the original configuration, this process repeats with new random regions until it identifies a candidate satisfying

$$\frac{FOM_0 - FOM_t}{FOM_0} > G_{th} \quad (2)$$

where G_{th} is the global searching threshold. The processes illustrated in Figure 1b are termed the global search phase, and the qualified region **B** is referred to as the global optimization area.

As demonstrated in Figure 2a, once a circular region **B** meets the specified condition (2), it is condensed to a smaller concentric circular region **S** (in this case, the radius of region **S** is 40 nm), referred to as the local optimization area. Within this region, the materials are permanently inverted, and FOM_0 is updated to reflect the current *FOM* value. Thereafter, the algorithm randomly selects a small adjacent region **X** next to **S** (Figure 2b), inverts the materials in **X**, calculates the trial value FOM_t, and then restores the materials in **X**. Repeat the above procedures until the following condition is met:

$$\frac{FOM_0 - FOM_t}{FOM_0} > L_{th} \tag{3}$$

where L_{th} is the local refinement threshold. When the region **X** satisfies the aforementioned condition (3), it is integrated into **S** (i.e., **S** = **S**∪**X**) (Figure 2c), subsequent to a permanent inversion of the materials in **X** and an update of FOM_0. The above procedures occurring in the vicinity of **S** are referred to as the local refinement phase, and the local optimization area **S** has the potential to expand through successive local searches until no adjacent region X meets the specified condition (Figure 2d). Once this occurs, the local refinement terminates, and a new cycle of the global search initiates (Figure 3a). The algorithm alternates between the global search and local refinement (Figure 3a–c) until FOM_0 achieves a value below a predefined performance threshold (FOM_s) or the cumulative number of search steps executed during either the global search or local refinement phases exceeds the preset maximum iteration limit (Figure 3d).

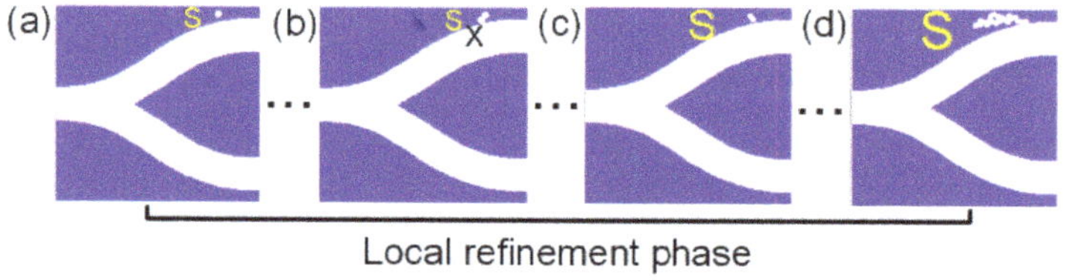

Figure 2. The local refinement phase. (**a**) Contract the global optimization region **B** satisfying condition (2) into a local optimization region **S**; (**b**) search for region **X** satisfying condition (3); and (**c**,**d**) region **X** satisfying condition (3) is progressively integrated into region **S**.

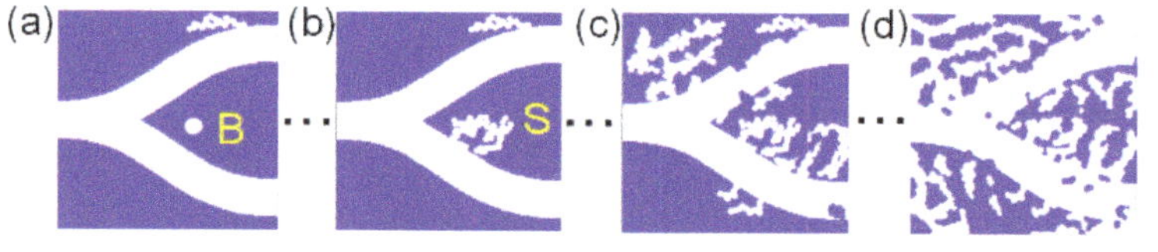

Figure 3. (**a**) The completion of the local refinement phase and the reinitiation of the global search phase; (**b**) the conclusion of the second local refinement phase; (**c**) the cyclic execution of global search and local refinement phases; and (**d**) the final optimized structure layout.

It is critical to emphasize that within the GLINT algorithm, the dimensions of regions **B** and **S** are user-configurable parameters that can either remain fixed or dynamically adapt during optimization, depending on specific optimization requirements. Let R_B and R_S denote the radii of regions **B** and **S**, respectively. Upon the completion of each local refinement phase and before recommencing the global search procedure, both parameters (R_B and R_S) can be adaptively adjusted based on the optimization progress. While a reduction in the size of region **B** inherently increases the global search duration, this strategy becomes essential when larger initial **B** regions fail to identify domains satisfying the global search threshold G_{th}. Under such conditions, strategically downsizing region **B** enables an accelerated convergence toward G_{th}-compliant regions, thereby significantly enhancing the algorithm's search efficiency in complex optimization scenarios. Similarly, G_{th} and L_{th} serve as adjustable parameters in the GLINT framework, and when tuning R_B and R_S in the algorithm, G_{th} and L_{th} can be simultaneously adjusted. In the presented example, these thresholds were empirically initialized at $G_{th} = 0.01$ and $L_{th} = 0.001$. The precise adjustment of G_{th}, L_{th}, and the dimensions of regions **B** and **S** is critical for identifying higher-quality local optima throughout the optimization process. To prevent the algorithm from stagnating due to a premature convergence to suboptimal local optima, we recommend employing larger initial parameters ($1 > G_{th} \geq 0.01$; $G_{th} \geq L_{th} \geq 0.001$; $R_B \geq 120$ nm; $R_B \geq R_S \geq 60$ nm) and progressively scaling them down throughout iterations. This adaptive strategy ensures balanced exploration–exploitation trade-offs, coordinating the rates of the search efficiency and the *FOM* reduction across the optimization trajectory. Notably, the selection of G_{th} and L_{th} retains full flexibility. Practitioners may alternatively initialize smaller optimization thresholds (e.g., $G_{th} = 1 \times 10^{-5}$, $L_{th} = 1 \times 10^{-7}$) along with reduced **B** and **S** dimensions (e.g., $R_B = 20$ nm, $R_S = 10$ nm) to prioritize the rapid acquisition of functionally viable devices, albeit with an increased risk of premature convergence to suboptimal local optima. This configurability empowers users to tailor the optimization workflow to application-specific priorities.

The GLINT inverse design algorithm introduces two fundamental advancements over the conventional DBS algorithm: First, it employs a dual-phase strategy integrating the global search with the local refinement, achieving a minimum feature size comparable to DTO (20 nm × 20 nm). This high-resolution capability enables the full exploitation of the optimization potential within the design space, directly resolving the limited flexibility issue inherent in the pixel-based DBS algorithm. Second, the global–local optimization architecture significantly enhances the computational efficiency through the rapid identification of critical optimization regions during the global phase, thereby addressing the computational inefficiency inherent in conventional DBS algorithms. Collectively, these innovations expand the applicability of the GLINT inverse design to complex photonic devices while simultaneously improving the computational efficiency and design flexibility. In the subsequent section, the practicability of our algorithm will be demonstrated by several examples.

3. Simulation and Experimental Results

3.1. Dual-Port WDM Structure

The complete simulation flow of the 1528.8/1607.5 nm WDM structure's crosstalk–iteration relationship (Section 2 case study) is presented in Figure 4a. Notably, this case maintains fixed R_B and R_S dimensions, with G_{th} and L_{th} adjusted only when algorithmic stagnation occurs due to undiscoverable regions satisfying conditions (2) and (3)—a deliberate configuration to benchmark the worst-case optimization speed. Utilizing a single i9-10900K CPU, the total optimization required 240 h across 1360 iterations. The slower optimization speed of this WDM structure is primarily attributed to the progressively diminishing probability of discovering region **B** configurations that satisfy the *FOM* reduction threshold during later optimization stages, where global search operations demanded excessive computational resources. Although reducing R_B/R_S would accelerate the search velocity in regions **B**/**S** and facilitate the escape from suboptimal local optima, these parameters were intentionally kept fixed in this case to validate the worst-case optimization performance.

Within Figure 4a, intermediate structures at iterations 1110 and 1215 exemplify GLINT's progressive optimization deceleration. At iteration 1110 (62 h), both ports' crosstalk reached −11.7 dB. By iteration 1215 (110 h), the crosstalk improved to −13.5 dB (upper port) and −11.7 dB (lower port). The markedly slower progression between iterations 1110 and 1215 indicates approaching optimization bottlenecks near local optima. Despite this, we continued reducing G_{th}/L_{th} thresholds to evaluate the performance characteristics of this local optimum. Upon reaching 240 h, the iteration rate degraded to approximately 1 step/3 h without the R_B/R_S adjustment, confirming the proximity to a local optimum. Final crosstalk values of −14.4 dB (upper port) and −12.5 dB (lower port) represent marginal improvements (<1 dB for both ports) over the 1215-iteration solution despite 130 additional hours. We emphasize that this configuration represents a near-optimal local solution (not global optimum) obtained within 110 h without parameter precise tuning. The precise co-optimization of R_B, R_S, G_{th}, and L_{th} would yield faster searching and superior local optima approaching the global optimum performance.

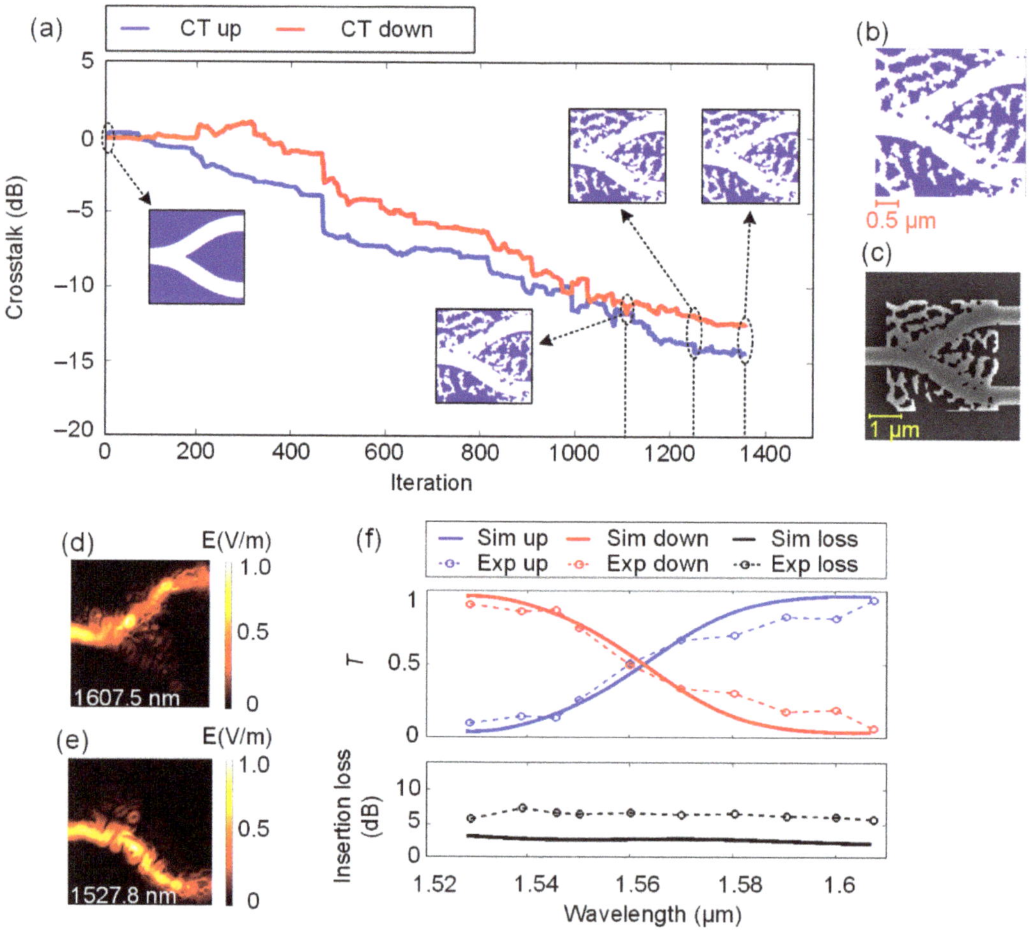

Figure 4. Experimental and simulation results for dual-port WDM structure. (**a**) Iteration steps and crosstalk schematic, and structures shown from left to right correspond to iterations 0, 1110, 1215, and 1360, respectively; (**b**) optimized structure layout; (**c**) SEM image of dual-ports WDM structure; (**d**,**e**) simulated electric field with 1607.5 nm and 1527.8 nm input; and (**f**) comparison of simulation and measurement results, T: normalized transmission ratio, T_i is defined as $T_i = P_i / \sum_k P_k$, where i, k ∈(up, down) and P_i are the output power at port i.

The schematic of the final optimized structure is shown in Figure 4b. The electric field distributions for input signals at wavelengths of 1607.5 nm and 1528.8 nm are depicted in Figures 4d and 4e, respectively, clearly demonstrating that the energy at different wavelengths is directed to different output ports. The upper port achieves a −14.4 dB crosstalk at 1607.5 nm, while the lower port provides a −12.5 dB crosstalk at 1528.8 nm, as shown in Figure 4f. To enable the systematic performance evaluation of the WDM structure, we define the normalized transmission ratio, denoted as T. More specifically, T_i designates this metric for the i-th output port, expressed as $T_i = P_i/\sum_k P_k$, where $i, k \in$ (up, down) and P_i are the output power at port i.

All nanostructures presented herein were fabricated via industry-standard electron beam lithography (EBL) with subsequent inductively coupled plasma (ICP) dry etching and wet chemical stripping, achieving minimum feature widths of 20 nm (±10 nm tolerance). The designed structure was fabricated on an SOI wafer. In the experimental setup, the output from a tunable laser source is directed through a polarization controller and into a lensed fiber, which subsequently couples the light to the face of the test chip. The output is similarly collected using a lensed fiber and measured by a power meter. To enhance the optical coupling efficiency at the chip's end face, inverse taper structures were introduced. Figure 4c shows an SEM image of the fabricated structure. It can be seen that several fine features present in Figure 4b are missing in Figure 4c. These deviations result in experimental normalized transmission ratio measurements (indicated by the red and blue dashed lines in Figure 4f) that do not closely match those simulation results, leading to an increased crosstalk of −9.7 dB at the upper port and −11.3 dB at the lower port, respectively. Similarly, the insertion loss (black dashed lines in Figure 4f) has also increased compared to the simulation results. These fine structural features originate from interstices between discrete circular refinement zones during the local optimization phase. When present in excessive quantities, such non-manufacturable features (<40 nm) significantly enlarge the discrepancy between simulated and experimental results. In our future work, these fine structural features will be effectively removed through spatial filtering, a heuristic learning-driven structural simplification, or an integrated LST algorithm to significantly enhance the device manufacturability.

3.2. The 90-Degree Bending Waveguides

The radii of conventional 90-degree bending waveguides typically range from the tens of micrometers, primarily due to the constraints imposed by radiation loss, which significantly hampers the development of ultra-compact optical integration. By employing the GLINT inverse design algorithm, the size of the 90-degree bending waveguide can be reduced by more than an order of magnitude. We validate the feasibility of the GLINT framework through the design, fabrication, and experimental characterization of a subwavelength-scale 90-degree bending waveguide, with a primary focus on demonstrating extreme dimensional miniaturization.

During the design process, a Euler bend with a radius of 1.4 μm is used as the initial structure, as illustrated in Figure 5a. The white regions represent silicon, while the blue regions denote silicon dioxide. The dimensions of the entire optimization region are 2 μm × 2 μm × 220 nm. After 11 h of iterative optimization, the final optimized structure is given in Figure 5b, with a 1.4 μm radius. Assuming an optical signal enters from the left side and exits from the top, the electric field distributions for the optimized structures are shown in Figure 5c.

The optimized structures were fabricated on a silicon-on-insulator (SOI) wafer. Figure 5d shows the scanning electron microscope (SEM) image of the test chip. The complete test structure contains twelve cascaded optimized 90-degree bending waveguides, only eight are visible here due to the limited imaging field of view. To mitigate the impact of fabrication errors, the average insertion loss of the individual waveguide structure is calculated and used as the experimental result.

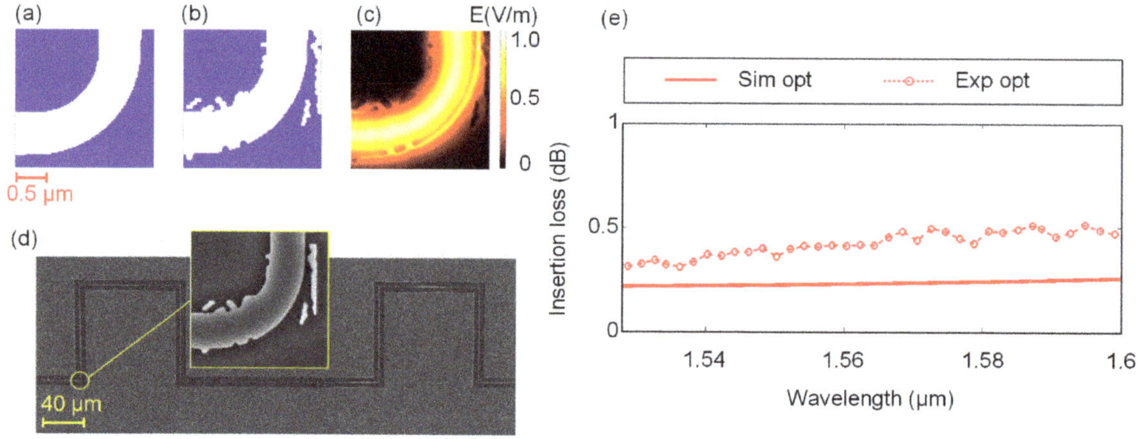

Figure 5. Experimental and simulation results for 90-degree bending waveguides. (**a**) Initial structure layout; (**b**) optimized structure layout; (**c**) electric field distribution of optimized structure; (**d**) SEM diagram on test chip; and (**e**) simulation and measurement results of insertion loss.

Figure 5e presents a comparison between the simulation and experimental results of the optimized structure. The solid line represents the simulation results, while the dashed line corresponds to the experimental data. The optimized structure achieves insertion losses as low as 0.22–0.26 dB across the 1528–1600 nm wavelength range in the simulation. Experimental measurements reveal insertion losses of 0.31–0.52 dB for the structure over the identical wavelength range. A comparative analysis of experimental and simulated results reveals an insertion loss increase of 0.09–0.26 dB in measured devices. This discrepancy is attributed to fabrication-induced losses; fabrication errors cause the structural deformation in the optimized configuration, deviating from the local optimum and consequently increasing the insertion loss.

3.3. Three-Port WDM Structure

To further evaluate the capability of our algorithm, we designed a more complex three-port WDM capable of separating input signals at wavelengths of 1450 nm, 1550 nm, and 1650 nm. As shown in Figure 6a, the optimized design measures 4.5 μm × 4.5 μm on a 220 nm thick SOI wafer. The input port is positioned on the left side, while the three output ports are located on the right side, arranged top to bottom.

The optimized structure exhibits insertion losses of 1.36 dB, 1.15 dB, and 1.08 dB at wavelengths of 1450 nm, 1550 nm, and 1650 nm, respectively, as shown in the lower portion of Figure 6b. Assuming the output powers for the upper, middle, and lower output ports are P_{up}, P_{middle}, and P_{down}, respectively, the normalized transmission ratio T_i is defined as $T_i = P_i / \sum_k P_k$, where $i, k \in$ (up, middle, down). T_{up}, T_{middle}, and T_{down} are also plotted in the upper portion of Figure 6b, represented by the red, green, and blue curves, respectively. Based on the electric field distribution simulation results shown in Figure 6c–e, at wavelengths of 1650 nm, 1550 nm, and 1450 nm, the input energy is effectively routed to the corresponding upper, middle, and lower ports, respectively. The crosstalk for each output port at three wavelengths is detailed in Table 1, with values ranging from −18.0 dB to −36.5 dB.

Throughout the optimization process, we dynamically scaled down R_{B}, R_{S}, G_{th}, and L_{th} parameters to prevent algorithmic stagnation. The parameter co-regulation strategy enabled the completion of a functionally more complex three-port WDM design within 190 h over 1842 iterations using a single CPU. Notably, despite doubling the simulation domain area compared to the dual-port WDM in Section 3.1, this three-port WDM design achieved both a faster optimization time and lower crosstalk—convincingly demonstrating that precise parameter co-regulation enhances GLINT's ability to escape suboptimal local optima and the optimization speed.

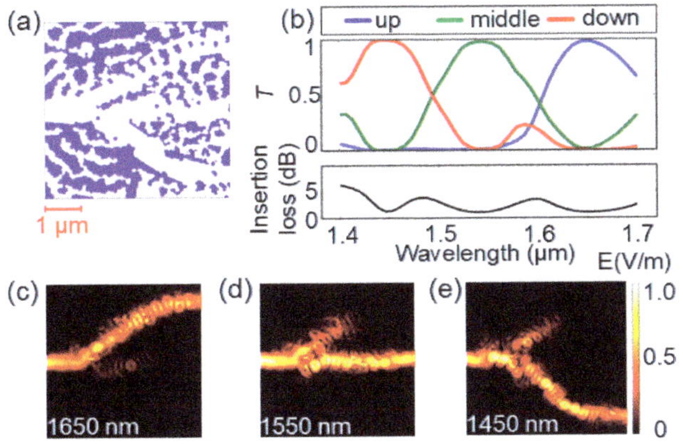

Figure 6. Simulation results for three-port WDM structure. (**a**) Optimized structure layout; (**b**) simulated normalized transmission ratio and insertion loss results; and (**c**–**e**) simulated electric field distribution with 1650 nm, 1550 nm, and 1450 nm input.

Table 1. Crosstalk simulation results for the three-port WDM at up, middle, and down ports.

Crosstalk (dB)		**Input Wavelength (μm)**		
		1.45	**1.55**	**1.65**
Output ports	up	−22.9	−18.1	—
	middle	−36.5	—	−18.0
	down	—	−20.1	−22.3

The compact size and excellent performance of this complex WDM structure demonstrate the robustness of our algorithm in addressing intricate design challenges. In future work, we will expand the simulation bandwidth of this WDM structure to enhance its tolerance to fabrication errors [28]. By simply modifying the optimization objective function, the design can be readily extended to accommodate more output ports. These compact WDMs have the potential to serve as key components in optical communication systems [29], optical neural networks [30], and optical frequency synthesizers [31,32].

4. Conclusions

In summary, we proposed the GLINT inverse design algorithm for integrated photonics, leveraging global search and local refinement strategies. The proposed algorithm addresses two critical limitations of conventional DBS algorithms through its unique global–local refinement framework. The high-precision local refinement phase resolves the constrained design freedom issue, while the efficient global search mechanism significantly enhances the computational efficiency. Compared to traditional DBS algorithms, this framework demonstrates a superior computational efficiency and design flexibility.

To demonstrate the effectiveness of the proposed algorithm, we have designed and experimentally validated a 90-degree bending waveguide structure and multiple WDM structures. The experimentally measured insertion loss of the optimized 90-degree bending waveguide ranges from 0.31 dB to 0.52 dB across the 1528–1600 nm wavelength. Additionally, the fabricated dual-port WDM structure achieves crosstalk values of −9.7 dB and −11.3 dB at the two output ports for the targeted wavelengths. The three-port WDM structure further demonstrates the algorithm's capability to optimize more complex design tasks.

Currently, we are actively extending our algorithm to further demonstrate its superior capabilities compared to conventional DBS algorithms. We are also developing several acceleration strategies to improve the algorithm's search speed and efficiency. In addition, we are exploring various heuristic methods that can regenerate the initial structure based on phased optimization results, thus simplifying the final optimized structure and improving its manufacturability. With ongoing advancements in computational power and fabrication precision, we anticipate that the GLINT inverse design algorithm will play an increasingly pivotal role in integrated photonics in the near future.

Author Contributions: Conceptualization, M.X. and M.L.; methodology, M.X.; software, M.X.; validation, M.L., T.W., Y.Z., Y.S., J.Y., D.P. and K.Z.; formal analysis, M.L.; investigation, M.L.; resources, M.L.; data curation, M.L.; writing—original draft preparation, M.L.; writing—review and editing, M.X.; visualization, M.L.; supervision, M.X.; project administration, M.X.; funding acquisition, M.X. All authors have read and agreed to the published version of the manuscript.

Funding: This work was supported by the National Key R&D Program of China (Grant No. 2021YFC2201902) and the National Natural Science Foundation of China (Grant No. 61975149).

Institutional Review Board Statement: Not applicable.

Informed Consent Statement: Not applicable.

Data Availability Statement: The authors will supply the relevant data in response to reasonable requests.

Conflicts of Interest: The authors declare no conflicts of interest.

References

1. Zhang, X.; Zhou, Z.; Guo, Y.; Zhuang, M.; Jin, W.; Shen, B.; Chen, Y.; Huang, J.; Tao, Z.; Jin, M. High-coherence parallelization in integrated photonics. *Nat. Commun.* **2024**, *15*, 7892. [CrossRef] [PubMed]
2. Weinstein, A.J.; Reed, M.D.; Jones, A.M.; Andrews, R.W.; Barnes, D.; Blumoff, J.Z.; Euliss, L.E.; Eng, K.; Fong, B.H.; Ha, S.D. Universal logic with encoded spin qubits in silicon. *Nature* **2023**, *615*, 817–822. [CrossRef] [PubMed]
3. Altug, H.; Oh, S.-H.; Maier, S.A.; Homola, J. Advances and applications of nanophotonic biosensors. *Nat. Nanotechnol.* **2022**, *17*, 5–16. [CrossRef] [PubMed]
4. Molesky, S.; Lin, Z.; Piggott, A.Y.; Jin, W.; Vucković, J.; Rodriguez, A.W. Inverse design in nanophotonics. *Nat. Photonics* **2018**, *12*, 659–670. [CrossRef]
5. Su, L.; Piggott, A.Y.; Sapra, N.V.; Petykiewicz, J.; Vuckovic, J. Inverse design and demonstration of a compact on-chip narrowband three-channel wavelength demultiplexer. *ACS Photonics* **2018**, *5*, 301–305. [CrossRef]
6. Lucas, E.; Yu, S.-P.; Briles, T.C.; Carlson, D.R.; Papp, S.B. Tailoring microcombs with inverse-designed, meta-dispersion microresonators. *Nat. Photonics* **2023**, *17*, 943–950. [CrossRef]
7. Yang, K.Y.; Skarda, J.; Cotrufo, M.; Dutt, A.; Ahn, G.H.; Sawaby, M.; Vercruysse, D.; Arbabian, A.; Fan, S.; Alù, A. Inverse-designed non-reciprocal pulse router for chip-based LiDAR. *Nat. Photonics* **2020**, *14*, 369–374. [CrossRef]
8. Liao, J.; Tian, Y.; Yang, Z.; Xu, H.; Tang, C.; Wang, Y.; Zhang, X.; Kang, Z. Inverse design of highly efficient and broadband mode splitter on SOI platform. *Chin. Opt. Lett.* **2024**, *22*, 011302. [CrossRef]
9. Yu, Z.; Cui, H.; Sun, X. Genetically optimized on-chip wideband ultracompact reflectors and Fabry–Perot cavities. *Photonics Res.* **2017**, *5*, B15–B19. [CrossRef]
10. Jin, X.; Xu, J.; Li, Y.; Xue, C.; Liao, R.; Fu, L.; Liu, M.; Shen, Y.; Quan, X.; Cheng, X. Fabrication-constrained inverse design and demonstration of high-performance grating couplers. *Chin. Opt. Lett.* **2024**, *22*, 112201. [CrossRef]
11. Yang, K.Y.; Shirpurkar, C.; White, A.D.; Zang, J.; Chang, L.; Ashtiani, F.; Guidry, M.A.; Lukin, D.M.; Pericherla, S.V.; Yang, J. Multi-dimensional data transmission using inverse-designed silicon photonics and microcombs. *Nat. Commun.* **2022**, *13*, 7862. [CrossRef] [PubMed]

12. Nikkhah, V.; Mencagli, M.J.; Engheta, N. Reconfigurable nonlinear optical element using tunable couplers and inverse-designed structure. *Nanophotonics* **2023**, *12*, 3019–3027. [CrossRef] [PubMed]
13. Hughes, T.W.; Minkov, M.; Williamson, I.A.; Fan, S. Adjoint method and inverse design for nonlinear nanophotonic devices. *ACS Photonics* **2018**, *5*, 4781–4787. [CrossRef]
14. Bendsøe, M.P. Optimal shape design as a material distribution problem. *Struct. Optim.* **1989**, *1*, 193–202. [CrossRef]
15. Frellsen, L.F.; Ding, Y.; Sigmund, O.; Frandsen, L.H. Topology optimized mode multiplexing in silicon-on-insulator photonic wire waveguides. *Opt. Express* **2016**, *24*, 16866–16873. [CrossRef] [PubMed]
16. Hammond, A.M.; Oskooi, A.; Johnson, S.G.; Ralph, S.E. Photonic topology optimization with semiconductor-foundry design-rule constraints. *Opt. Express* **2021**, *29*, 23916–23938. [CrossRef] [PubMed]
17. Lalau-Keraly, C.M.; Bhargava, S.; Miller, O.D.; Yablonovitch, E. Adjoint shape optimization applied to electromagnetic design. *Opt. Express* **2013**, *21*, 21693–21701. [CrossRef] [PubMed]
18. Piggott, A.Y.; Petykiewicz, J.; Su, L.; Vučković, J. Fabrication-constrained nanophotonic inverse design. *Sci. Rep.* **2017**, *7*, 1786. [CrossRef] [PubMed]
19. Piggott, A.Y.; Lu, J.; Lagoudakis, K.G.; Petykiewicz, J.; Babinec, T.M.; Vučković, J. Inverse design and demonstration of a compact and broadband on-chip wavelength demultiplexer. *Nat. Photonics* **2015**, *9*, 374–377. [CrossRef]
20. Shen, B.; Wang, P.; Polson, R.; Menon, R. An integrated-nanophotonics polarization beamsplitter with $2.4 \times 2.4\ \mu m^2$ footprint. *Nat. Photonics* **2015**, *9*, 378–382. [CrossRef]
21. Shen, B.; Wang, P.; Polson, R.; Menon, R. Integrated metamaterials for efficient and compact free-space-to-waveguide coupling. *Opt. Express* **2014**, *22*, 27175–27182. [CrossRef] [PubMed]
22. Li, Z.; Li, G.; Huang, J.; Zhang, Z.; Yang, J.; Yang, C.; Qian, Y.; Xu, W.; Huang, H. Ultra-compact high efficiency and low crosstalk optical interconnection structures based on inverse designed nanophotonic elements. *Sci. Rep.* **2020**, *10*, 11993. [CrossRef] [PubMed]
23. Sebbag, Y.; Talker, E.; Naiman, A.; Barash, Y.; Levy, U. Demonstration of an integrated nanophotonic chip-scale alkali vapor magnetometer using inverse design. *Light Sci. Appl.* **2021**, *10*, 54. [CrossRef] [PubMed]
24. Liu, Y.; Li, H.; Chen, W.; Wang, P.; Dai, S.; Zhang, B.; Li, J.; Li, Y.; Fu, Q.; Dai, T. Direct-binary-search-optimized compact silicon-based polarization beam splitter using a pixelated directional coupler. *Opt. Commun.* **2021**, *484*, 126670. [CrossRef]
25. Xu, K.; Liu, L.; Wen, X.; Sun, W.; Zhang, N.; Yi, N.; Sun, S.; Xiao, S.; Song, Q. Integrated photonic power divider with arbitrary power ratios. *Opt. Lett.* **2017**, *42*, 855–858. [CrossRef] [PubMed]
26. Lu, L.; Liu, D.; Zhou, F.; Li, D.; Cheng, M.; Deng, L.; Fu, S.; Xia, J.; Zhang, M. Inverse-designed single-step-etched colorless 3 dB couplers based on RIE-lag-insensitive PhC-like subwavelength structures. *Opt. Lett.* **2016**, *41*, 5051–5054. [CrossRef] [PubMed]
27. Ma, H.; Huang, J.; Zhang, K.; Yang, J. Ultra-compact and efficient 1×2 mode converters based on rotatable direct-binary-search algorithm. *Opt. Express* **2020**, *28*, 17010–17019. [CrossRef] [PubMed]
28. Lu, J.; Vučković, J. Nanophotonic computational design. *Opt. Express* **2013**, *21*, 13351–13367. [CrossRef] [PubMed]
29. Li, N.; Xin, M.; Su, Z.; Magden, E.S.; Singh, N.; Notaros, J.; Timurdogan, E.; Purnawirman, P.; Bradley, J.D.; Watts, M.R. A silicon photonic data link with a monolithic erbium-doped laser. *Sci. Rep.* **2020**, *10*, 1114. [CrossRef] [PubMed]
30. Xu, X.; Tan, M.; Corcoran, B.; Wu, J.; Boes, A.; Nguyen, T.G.; Chu, S.T.; Little, B.E.; Hicks, D.G.; Morandotti, R. 11 TOPS photonic convolutional accelerator for optical neural networks. *Nature* **2021**, *589*, 44–51. [CrossRef] [PubMed]
31. Singh, N.; Xin, M.; Li, N.; Vermeulen, D.; Ruocco, A.; Magden, E.S.; Shtyrkova, K.; Ippen, E.; Kärtner, F.X.; Watts, M.R. Silicon photonics optical frequency synthesizer. *Laser Photonics Rev.* **2020**, *14*, 1900449. [CrossRef]
32. Xin, M.; Li, N.; Singh, N.; Ruocco, A.; Su, Z.; Magden, E.S.; Notaros, J.; Vermeulen, D.; Ippen, E.P.; Watts, M.R. Optical frequency synthesizer with an integrated erbium tunable laser. *Light Sci. Appl.* **2019**, *8*, 122. [CrossRef] [PubMed]

photonics

MDPI

Article

Ultra-Compact Inverse-Designed Integrated Photonic Matrix Compute Core

Mingzhe Li [1], Tong Wang [1], Yi Zhang [1], Yulin Shen [1], Jie Yang [1], Ke Zhang [1], Dehui Pan [1], Jiahui Yao [1] and Ming Xin [1,2,*]

[1] School of Electrical and Information Engineering, Tianjin University, Tianjin 300072, China; lmz804472764@tju.edu.cn (M.L.); wttesp@tju.edu.cn (T.W.); zy_1@tju.edu.cn (Y.Z.); shenyl@tju.edu.cn (Y.S.); yangjiewg@tju.edu.cn (J.Y.); zhangke2672@tju.edu.cn (K.Z.); panda@tju.edu.cn (D.P.); yaojiahui@tju.edu.cn (J.Y.)

[2] Tianjin Key Laboratory of Brain-Inspired Intelligence Technology, Tianjin 300072, China

* Correspondence: xinm@tju.edu.cn

Abstract

Leveraging our developed Global–Local Integrated Topology inverse design algorithm, we designed an efficient, compact, and symmetrical power splitter on a silicon-on-insulator platform. This device achieves a low insertion loss of 0.18 dB and a power imbalance of <0.0002 dB between its output ports within an ultra-compact footprint of 5.5 μm × 2.5 μm. The splitter, combined with an ultra-compact 0–π phase shifter measuring only 4.5 μm × 0.9 μm on the silicon-on-insulator platform, forms an ultra-compact inverse-designed integrated photonic matrix compute core, thus enabling the function of matrix operations in optical neural networks. Through a networked cascade of power splitters and phase shifters, this silicon-based photonic matrix compute core achieves an integration density of ~26,000 computational units/mm^2. Moreover, it attained 99.05% accuracy in handwritten digit recognition (0–9) and exhibited strong robustness against fabrication errors, maintaining >80% accuracy with >0.9 probability under simulated random fabrication errors.

Keywords: integrated photonic devices; inverse design; optical neural network

1. Introduction

The rapid advancement of semiconductor technology has enabled artificial neural networks (ANNs) to achieve remarkable success in domains including object classification [1], computer vision [2], real-time translation [3], and autonomous driving [4]. However, conventional ANNs exhibit computational speed limitations intrinsic to the von Neumann architecture [5]. Within this context, optical neural networks (ONNs) have emerged as promising alternatives to traditional ANNs owing to their inherent advantages: ultra-wide-bandwidth, high computational speed, and massive parallelism [6–8]. Existing ONN implementations include Mach–Zehnder interferometer (MZI) cascades [6,9–11], integrated diffractive optical neural networks (DONNs) [12–15], compact multi-mode interference convolutional processors [16], and neural networks implemented using phase-change materials (PCMs) [17]. Among these approaches, ONNs based on the multistage interference principle, such as MZI-based ONNs, have been widely adopted by researchers due to their high scalability and excellent classification accuracy [18]. However, these methods remain constrained by low integration density (<10^3 units/mm^2, typically ~10^2

units/mm^2). Constructing highly-integrated photonic platforms remains a significant challenge for multistage interference-based ONNs.

In recent years, inverse design methodologies have emerged as powerful tools for developing ultra-compact integrated photonic devices [19–26], significantly enhancing the integration density of photonic platforms. In our previous work, we developed the Global–Local Integrated Topology (GLINT) inverse design algorithm [27]—a global–local co-optimization framework that enables direct optimization of binary waveguide-silica structures. Leveraging the GLINT algorithm, we designed an ultra-compact, symmetric power splitter (5.5 μm × 2.5 μm) that achieves an insertion loss of 0.18 dB and a power imbalance below 0.0002 dB. Through network-level cascading of power splitters and novel compact phase shifters (4.5 μm × 0.9 μm), we constructed an ultra-compact inverse-designed integrated photonic matrix compute core (PMCC, 132 μm × 16 μm). The proposed PMCC, capable of performing matrix operations for ONNs, was evaluated on the Modified National Institute of Standards and Technology (MNIST) handwritten digit classification task, achieving a classification accuracy of 99.05%. Notably, our PMCC achieves an integration density of 2.6×10^4 computational units/mm^2—adopting a multistage interferometric architecture similar to that of MZI-based ONNs, yet yielding a significantly higher integration density.

Furthermore, to evaluate fabrication-error tolerance, we developed a stochastic fabrication-error model that incorporates over-etching, under-etching, and etch-induced deformations in inversely designed geometries. These errors were introduced as normally distributed random parameters. Using this model, we constructed 1000 PMCC structures, each with random fabrication errors, and evaluated each structure on 10,000 MNIST test images. Statistical analysis revealed that more than 90% of the simulation test results maintained greater than 80% recognition accuracy, demonstrating exceptional process robustness in our PMCC.

In the following sections, we first introduce the fundamental principles of the PMCC in Section 2, presenting its general architecture. Subsequently, Sections 3 and 4 detail the essential components of the PMCC: the compact symmetric power splitter and the compact phase shifter, respectively. Following this, Section 5 demonstrates an application example of the PMCC for handwritten digit recognition (0–9). Finally, Section 6 establishes a unique stochastic fabrication-error simulation model to validate the robustness of the system.

2. Introduction to the Principles of the PMCC

This section provides a concise introduction to the fundamental principles of the PMCC. Figure 1 illustrates the basic architecture of a PMCC with $2n$ input and output ports, which includes general optical locally-connected (OLC) layers, and provides a schematic overview of the operational process of the PMCC. Specifically, upon acquisition by edge devices, multimodal data such as video, audio, and images undergo feature preprocessing and are subsequently processed by the signal modulation layer (SM layer), which encodes the information onto the incident optical carriers before introducing them into the system. After the information-carrying light passes through n OLC layers—where complex all-optical matrix operations are performed on the feature vectors—the output signals are captured by a photodiode array (PD array), enabling functionalities such as real-time translation, autonomous driving, and image recognition.

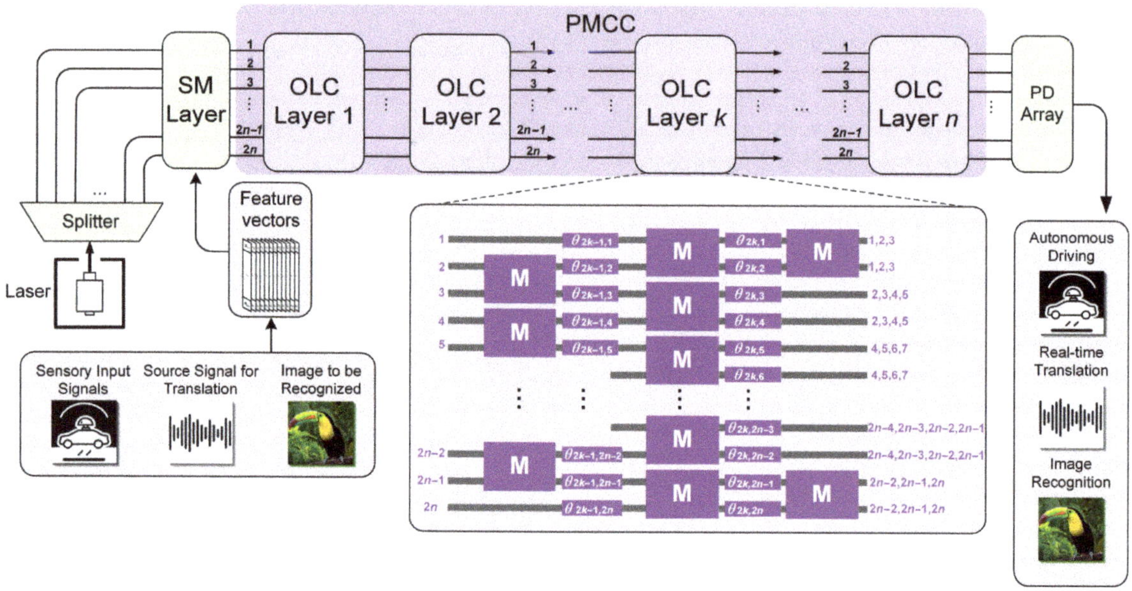

Figure 1. Schematic of the proposed PMCC. Abbreviations: SM layer, signal modulation layer; PD array, photodiode array.

The specific network connectivity of this OLC layer is shown in Figure 1. The function of the OLC layer is to establish partial connectivity between input and output ports and to impart an independent phase shift to the signal from each input port. Specifically, output ports 1, 2 and output ports $2n - 1$, $2n$ can be connected to input ports 1, 2, 3 and input ports $2n - 2$, $2n - 1$, $2n$, respectively. For output ports $2j - 1$ and $2j$ ($1 < j < n$), both can be connected to input ports $2j - 2$, $2j - 1$, $2j$, and $2j + 1$, thereby achieving partial interconnection between input and output ports. The PMCC is constructed by cascading n OLC layers, each consisting of $2n + 1$ 2×2 power splitters M and $4n$ phase shifters $\theta_{s,t}$ (s, t = 1, 2, …, $2n$). The input information is modulated in parallel onto $2n$ continuous-wave laser fields from the same laser source, which are then injected into the $2n$ input ports of the PMCC. After processing by the PMCC, the resulting $2n$ output optical fields can either be directed to subsequent optical processing modules or be demodulated and converted into electronic signals via photodetectors to retrieve the processed information. The transfer function of the phase shifter $\theta_{s,t}$ is given by $e^{i\theta_{s,t}}$, and that of the power splitter M is expressed as:

$$\mathbf{A} = \begin{bmatrix} A_{11} & A_{12} \\ A_{21} & A_{22} \end{bmatrix} \tag{1}$$

Under ideal conditions where the insertion loss of the power splitters is negligible, $\mathbf{A}$ is strictly a unitary matrix. According to the input-output connectivity rules of the OLC layer, after cascading through n OLC layers, the signal at each of the final $2n$ output ports corresponds to a weighted sum of the signals from the initial $2n$ input ports. To ensure that the energy from each input port is distributed as uniformly as possible across the different output ports, M should be a 50:50 power splitter satisfying:

$$|A_{pq}| = \frac{1}{\sqrt{2}} \tag{2}$$

where $p, q = 1, 2$. Furthermore, assume that $\mathbf{T}_1$, $\mathbf{T}_2$, $\mathbf{T}_3$, $\mathbf{P}_k$ ($k = 1, 2, \ldots, 2n$) are all $2n \times 2n$ matrices satisfying the following conditions:

$$\mathbf{T}_1 = \begin{bmatrix} 1 & 0 & \ldots & 0 & 0 \\ 0 & \mathbf{A} & \ldots & 0 & 0 \\ \vdots & \vdots & \ddots & \vdots & \vdots \\ 0 & 0 & \ldots & \mathbf{A} & 0 \\ 0 & 0 & \ldots & 0 & 1 \end{bmatrix} \quad (3)$$

$$\mathbf{T}_2 = \begin{bmatrix} \mathbf{A} & 0 & \ldots & 0 & 0 \\ 0 & \mathbf{A} & \ldots & 0 & 0 \\ \vdots & \vdots & \ddots & \vdots & \vdots \\ 0 & 0 & \ldots & \mathbf{A} & 0 \\ 0 & 0 & \ldots & 0 & \mathbf{A} \end{bmatrix} \quad (4)$$

$$\mathbf{T}_3 = \begin{bmatrix} \mathbf{A} & 0 & \ldots & 0 & 0 \\ 0 & 1 & \ldots & 0 & 0 \\ \vdots & \vdots & \ddots & \vdots & \vdots \\ 0 & 0 & \ldots & 1 & 0 \\ 0 & 0 & \ldots & 0 & \mathbf{A} \end{bmatrix} \quad (5)$$

$$\mathbf{P}_k = \begin{bmatrix} e^{i\theta_{k,1}} & 0 & \ldots & 0 & 0 \\ 0 & e^{i\theta_{k,2}} & \ldots & 0 & 0 \\ \vdots & \vdots & \ddots & \vdots & \vdots \\ 0 & 0 & \ldots & e^{i\theta_{k,2n-1}} & 0 \\ 0 & 0 & \ldots & 0 & e^{i\theta_{k,2n}} \end{bmatrix} \quad (6)$$

Then the transfer function of the entire PMCC is given by:

$$\mathbf{T}_{\mathrm{PMCC}} = \prod_{k=n}^{1} \mathbf{T}_3\mathbf{P}_{2k}\mathbf{T}_2\mathbf{P}_{2k-1}\mathbf{T}_1 \quad (7)$$

Based on the unitarity of $\mathbf{A}$, it follows that $\mathbf{T}_1$, $\mathbf{T}_2$, $\mathbf{T}_3$ are also unitary matrices, and thus $\mathbf{T}_{\mathrm{PMCC}}$ is likewise unitary. If the phase shift of each phase shifter is regarded as an independent variable of $\mathbf{T}_{\mathrm{PMCC}}$, then $\mathbf{T}_{\mathrm{PMCC}}$ possesses $4n^2$ independent variables, enabling its application in complex quantum computations and theoretical physics simulations [28].

3. Compact Symmetric Power Splitter

Compact power splitters represent the first essential component of the PMCC, designed using our GLINT inverse design algorithm. We first provide a concise overview of the GLINT algorithm [27] employed in designing this device. This algorithm employs a trajectory-based optimization strategy and iteratively modifies the structure by flipping material states within waveguide-substrate regions, enabling direct optimization of binary photonic structures. The GLINT algorithm comprises two distinct phases: global search and local refinement. The global search phase identifies performance-critical regions using large-scale global search regions, while local refinement utilizes small-scale local optimization regions to optimize features, thus achieving a 20 nm $\times$ 20 nm pixel size for optimization while maintaining computational efficiency. Device performance is quantified by a Figure of Merit (FOM, $0 \leq FOM \leq 1$), where lower values indicate superior performance. All simulation results during algorithm iterations were obtained via the three-dimensional finite-difference time-domain (3D-FDTD) method with a spatial resolution of 20 nm × 20 nm and with perfectly matched layer (PML) boundary conditions.

In the original global–local optimization framework, the center points for both global and local optimization regions are randomly selected within the optimization region. To reduce computational cost, we introduced a symmetry constraint module, which confines candidate center points for global optimization and local refinement to the upper half-region of the optimization region. During each waveguide-substrate material inversion operation, synchronized inversion is performed in the geometrically symmetric lower region, ensuring structural symmetry preservation throughout all optimization iterations.

Figure 2a shows the initial structure of the compact power splitter designed on a 220-nm silicon-on-insulator (SOI) platform with dimensions of 5.5 μm × 2.5 μm, where silicon is represented in white and silica in blue, featuring two input ports on the left and two output ports on the right. The width of the input and output waveguides is 0.5 μm, with 1.6 μm spacing between the upper and lower ports. In the absence of the symmetry constraint, simulations must be conducted separately for signals input through each of the two ports during every step of the optimization iteration. By incorporating the symmetry module, however, the simulation results for the upper input port can be symmetrically mapped to deduce that of the lower input port, thereby reducing the computational cost by half. Figure 2c presents the schematic of the optimized power splitter after inverse design. This symmetric compact device operates at 1550 nm wavelength. Figure 2b,d show the simulated electric field distributions in the initial and optimized structures, respectively, with a 1550 nm signal applied at the upper input port. To satisfy Equation (2), the *FOM* is defined as:

$$FOM = 1 - (1 - |0.5 - t_{11}|) \times (1 - |0.5 - t_{12}|) \quad (8)$$

here, t_{11} denotes the transmission at the upper output port when a 1550 nm signal is applied to the upper input port, and t_{12} represents the transmission at the lower output port under the same input conditions.

The initial structure employs a coupled-waveguide design principle, which utilizes evanescent field coupling between two closely spaced curved waveguides to achieve energy exchange. While achieving a 1:1 splitting ratio solely through coupling typically requires coupling lengths of tens to hundreds of microns, the GLINT algorithm optimizes the coupling region, enabling it to achieve the same functionality at a length of just 5.5 μm. Although this initial structure does not achieve the precise 1:1 splitting ratio—instead producing an approximate 1:10 ratio (Figure 2b)—its coupling characteristics establish the foundation for subsequent optimization. By optimizing the geometry of the optimization region, the GLINT algorithm enables precise control of the splitting ratio, which fulfills the target performance specifications of the power splitter. Figure 2e presents the transmission spectrum of the optimized structure under upper input port excitation, with blue/red curves representing transmission at the upper and lower output ports, respectively. The device exhibits 0.18 dB insertion loss at 1550 nm, where both output ports achieve approximately −3.19 dB transmission. The power imbalance (<0.0002 dB) at this wavelength is negligible—a finding consistent with the symmetric electric field distribution shown in Figure 2d. Furthermore, as indicated by the transmission curves in Figure 2e, the power imbalance of the splitter remains below 0.15 dB over an input wavelength range of 1550 ± 5 nm.

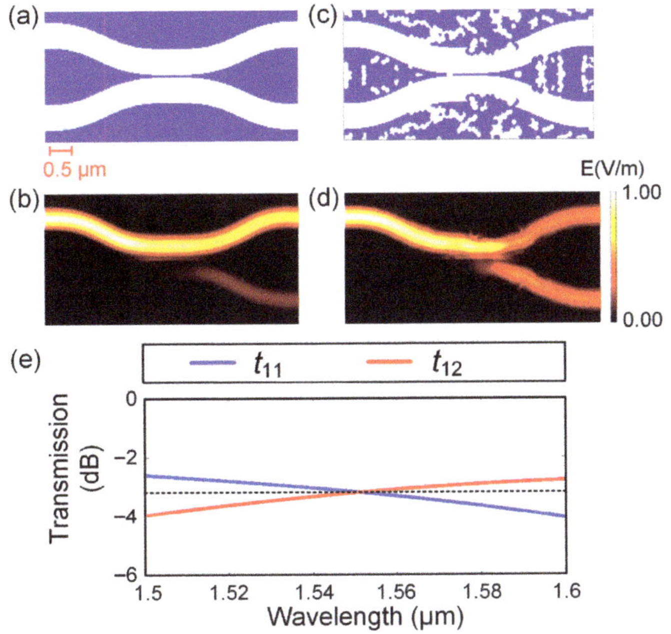

Figure 2. Compact symmetric power splitter. (**a**) Initial structure (white: silicon; blue: silica); (**b**) electric field distribution of the initial structure with 1550 nm input at the upper input port; (**c**) optimized 2 × 2 compact symmetric power splitter; (**d**) electric field distribution of the optimized power splitter with 1550 nm input at the upper input port; (**e**) transmission of the optimized structure. The dashed lines indicate the transmission at 1550 nm for the two output ports.

The electric field transmission matrix (transfer function) of the optimized structure can be expressed as:

$$\mathbf{A_E} = \begin{bmatrix} -0.657 + 0.219i & -0.211 - 0.660i \\ -0.211 - 0.660i & -0.657 + 0.219i \end{bmatrix} \tag{9}$$

Herein, the input/output ports are sequentially numbered from top to bottom as port 1 and 2. For the complex matrix $\mathbf{A_E}$, the element $A_E(i,j)$ represents the electric field transmission ratio from input port j to output port i (i, j = 1, 2) under 1550 nm wavelength. The power transmission matrix $\mathbf{A_P}$ is obtained by taking the squared modulus of each element in $\mathbf{A_E}$:

$$\mathbf{A_P} = \begin{bmatrix} 0.480057 & 0.480038 \\ 0.480038 & 0.480057 \end{bmatrix} \tag{10}$$

The $A_P(i,j)$ represents the transmission ratio from input port j to output port i (i, j = 1, 2) under 1550 nm wavelength. As evidenced by $\mathbf{A_P}$, the differential transmission between output ports is 0.0019%—approaching zero. This indicates exceptional power-splitting symmetry where optical power is equally divided between dual outputs with high precision. The high-performance design provides a buffering effect against fabrication errors, which significantly improves the PMCC's robustness.

Finally, the impact of temperature variations on the power splitter was evaluated. The thermo-optic coefficients of Si and SiO_2 at the operating wavelength are approximately 1.86×10^{-4} K^{-1} and 1.0×10^{-5} K^{-1}, respectively. Based on 3D-FDTD simulations, a temperature variation of ±5 °C results in a power variation of less than 0.6% and a phase shift below $6.4 \times 10^{-4}\pi$ at the output ports, demonstrating the robustness of the device against temperature fluctuations.

4. Compact Phase Shifter

In addition to power splitters, phase shifters represent another essential component for realizing multistage interference in PMCC. We further propose a compact phase shifter that offers a 0–π modulation range achieved by varying the width (W) of the structure.

Figure 3a shows the design structure of the phase shifter, with a maximum footprint of 4.5 μm × 0.9 μm. The core component of the phase shifter is a tunable region with a variable W, ranging from 0.5 μm to 0.9 μm. The widths of both the input and output waveguides are fixed at 0.5 μm. When the width W of the tunable region is 0.5 μm, the structure is equivalent to a straight waveguide. By adjusting the width W, the effective index within the modulation region can be modified, thereby achieving phase modulation. The electric field phase difference between the output and input ports for the straight waveguide case (W = 0.5 μm) is recorded as θ_0. After adjusting the width W, the simulated electric field phase difference between the output and input ports is recorded as θ_W. The phase shift θ introduced by the phase shifter is defined as:

$$\theta = \theta_W - \theta_0 \tag{11}$$

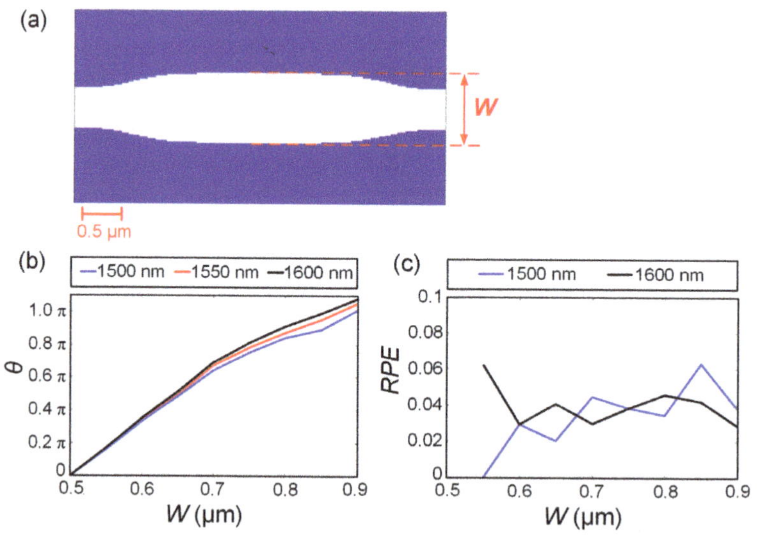

Figure 3. Compact phase shifter. (**a**) Simulated structure of the compact phase shifter (white: silicon; blue: silica); (**b**) relationship between structure width W and phase shift θ in the phase shifter; (**c**) relative phase error (*RPE*) of the phase shifter at 1500 nm and 1600 nm inputs, referenced to the 1550 nm result.

In essence, θ represents the additional phase shift introduced by the phase shifter relative to a straight waveguide of identical length.

Figure 3b depicts the relationship between the structure width W and the phase shift θ. Simulation results demonstrate that for W ranging from 0.5 μm to 0.9 μm under 1550 nm optical input, the structure achieves a maximum phase shift of π with virtually negligible insertion losses (<5.8×10^{-3} dB). A quasi-linear relationship exists between θ and W, which enables straightforward determination of the required W value for a target phase shift through simple calculation, thereby significantly streamlining the device design process. To evaluate the wavelength sensitivity of the structure, the relationship between θ and W was further calculated at input wavelengths of 1500 nm and 1600 nm, with the corresponding data represented by the blue and black curves in Figure 3b. Here, we introduce the concept of relative phase error (*RPE*)—defined as the relative error in phase shift when the input wavelength deviates from the design wavelength—to quantitatively evaluate wavelength sensitivity:

$$RPE = \left|\frac{\theta_\lambda - \theta_{1550}}{\theta_{1550}}\right| \tag{12}$$

where θ_λ denotes the phase shift θ at an input wavelength of λ (with λ = 1500 nm and 1600 nm corresponding to the blue and black curves in Figure 3b, respectively), and θ_{1550} represents the phase shift θ at the design wavelength of 1550 nm (red curve in Figure 3b). Figure 3c demonstrates that the percentage deviation in phase shift is less than 7% under a wavelength deviation of ±50 nm from the design wavelength. Furthermore, in the error analysis (Section 6), even after introducing both a maximum phase deviation of $\pi/20$ (corresponding to a mean *RPE* of 16%) in the phase shifters and the etching errors in the power splitters, our PMCC maintained strong performance in the handwritten digit recognition task, achieving over 80% accuracy in more than 90% of the trials. This robust performance indirectly demonstrates the limited impact of a ±50 nm wavelength deviation on the system.

Furthermore, we evaluated the impact of temperature variations on the phase shifter. Based on 3D-FDTD simulations, a temperature variation of ±5 °C induces a phase shift of less than 0.0019π radians, demonstrating the robustness of the device against temperature fluctuations.

5. Constructing the Compact Optical Neural Network

In this subsection, to accomplish the task of recognizing handwritten digits (0–9), we constructed a PMCC by network-level cascading of 55 power splitters and 100 phase shifters (Figure 4a). The resulting network comprises 10 input ports and 10 output ports, establishing full signal connectivity from all inputs to all outputs. This architecture enables arbitrary weight tuning for neural networks through phase adjustment in individual phase shifters. The proposed network is capable of performing 10 × 10 matrix operations. To calculate the integration density, the structure comprising one power splitter and two compact phase shifters (Figure 4b) is defined as a compact computational unit (CCU, 10 µm × 3 µm). Based on this definition, the implemented PMCC, with a footprint of 132 µm × 16 µm, achieves an integration density of approximately 26,000 units per mm^2.

Furthermore, to validate the layout feasibility of the designed ultra-compact PMCC, we performed 3D-FDTD simulations (with a resolution of 20 nm × 20 nm) on adjacent power splitters and phase shifters to quantify the crosstalk between adjacent devices. In the PMCC, the center-to-center spacing between vertically adjacent power splitters is 3.2 µm, the minimum input/output waveguide spacing is 1.6 µm, and the minimum spacing between adjacent phase shifters is only 0.7 µm. Simulation results demonstrate that the crosstalk between adjacent power splitters remains below −57 dB, while that between adjacent phase shifters is below −68 dB, both being extremely low and thereby confirming the rationality of the PMCC layout.

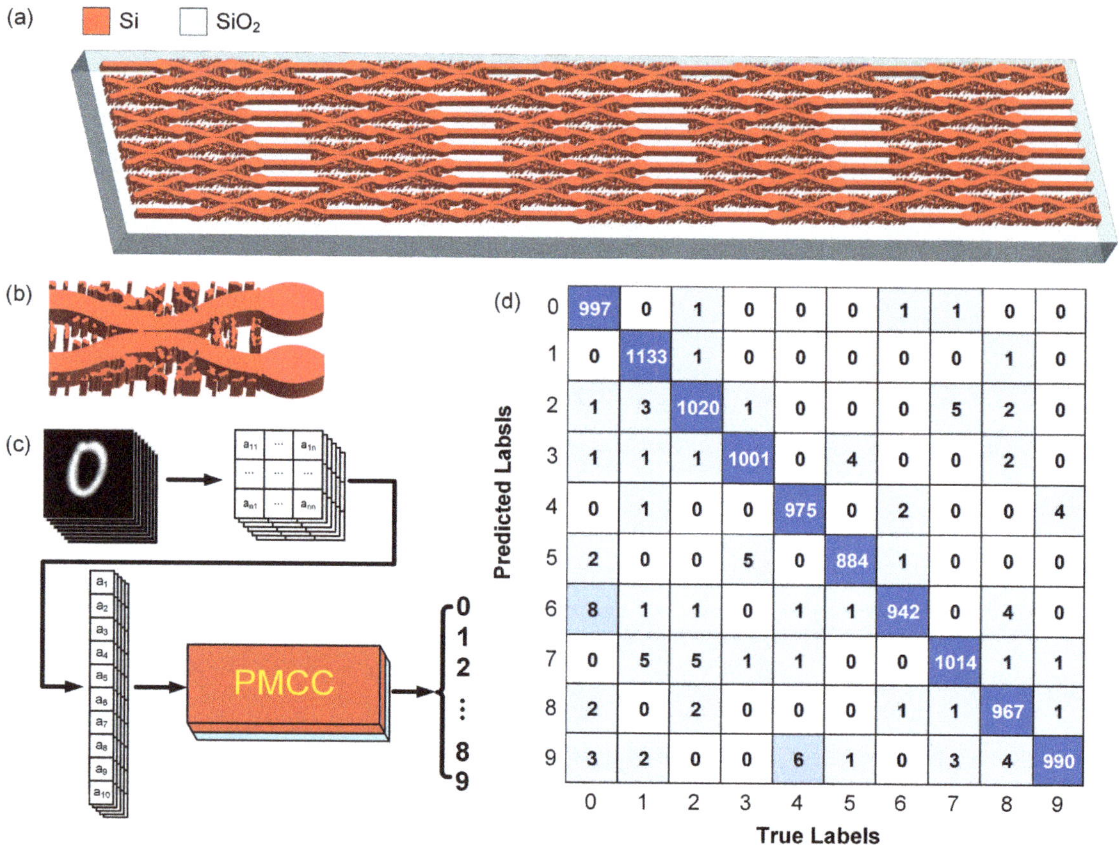

Figure 4. The compact linear matrix operator. (**a**) Architecture of the PMCC for 0–9 handwritten digit recognition, Si is denoted by red, and SiO_2 by gray; (**b**) structural illustration of the CCU; (**c**) schematic diagram of handwritten digit recognition using the PMCC; (**d**) confusion matrix obtained from simulated testing of the 0–9 handwritten digit classification task using 10,000 test samples after network training. (The SiO_2 cladding is omitted in (**a**) for clarity of illustration).

Herein, we employ the complex MNIST dataset to validate the functionality of the constructed PMCC (Figure 4c). This dataset comprises 60,000 training images and 10,000 test images, each being a 28 × 28-pixel grayscale image labeled across 10 digits classes (0–9). For each training image, the 28 × 28 pixels matrix is converted into a 10 × 1 feature vector via conventional neural network techniques. This feature vector is subsequently fed into the ten input ports on the left side of the PMCC. Training involves adjusting the phases of 100 phase shifters within the PMCC.

Prior to the initiation of training, the gradient of the loss function with respect to $\theta_{s,t}$ is computed. Within the general framework illustrated in Figure 1, the input electric field vector of the PMCC is defined as $\mathbf{X} = [X_1, X_2, \ldots, X_{2n}]^T$ (the 10 × 1 feature vector in this subsection), while the output electric field vector is denoted as $\mathbf{Y} = [Y_1, Y_2, \ldots, Y_{2n}]^T$. After photodetection, the output is given by $\mathbf{F} = [F_1, F_2, \ldots, F_{2n}]^T$, where $F_k = |Y_k|^2$ for $k = 1, 2, \ldots, 2n$. Denoting the loss function of the deep learning network as $L(\mathbf{F})$, the gradient of the loss function with respect to the phase shift parameter $\theta_{s,t}$ of the PMCC can be expressed as follows:

$$\frac{\partial L}{\partial \theta_{s,t}} = 2Re\left[ie^{i\theta_{s,t}}\sum_k \frac{\partial L}{\partial F_k} R^s_{k,t} Y^*_k \sum_l Q^s_{t,l} X_l\right] \tag{13}$$

where:

$$\mathbf{R}^s = \begin{cases} \left(\prod_{k=n}^{\frac{s}{2}+1} \mathbf{T}_3\mathbf{P}_{2k}\mathbf{T}_2\mathbf{P}_{2k-1}\mathbf{T}_1 \right) \mathbf{T}_3, & \mathrm{mod}(s,2) = 0 \\ \left(\prod_{k=n}^{\frac{(s+3)}{2}} \mathbf{T}_3\mathbf{P}_{2k}\mathbf{T}_2\mathbf{P}_{2k-1}\mathbf{T}_1 \right) \mathbf{T}_3\mathbf{P}_{s+1}\mathbf{T}_2, & \mathrm{mod}(s,2) = 1 \end{cases} \tag{14}$$

$$\mathbf{Q}^s = \begin{cases} \mathbf{T}_2\mathbf{P}_{s-1}\mathbf{T}_1 \left(\prod_{k=\frac{s}{2}-1}^{1} \mathbf{T}_3\mathbf{P}_{2k}\mathbf{T}_2\mathbf{P}_{2k-1}\mathbf{T}_1 \right), & \mathrm{mod}(s,2) = 0 \\ \mathbf{T}_1 \left(\prod_{k=\frac{(s-3)}{2}}^{1} \mathbf{T}_3\mathbf{P}_{2k}\mathbf{T}_2\mathbf{P}_{2k-1}\mathbf{T}_1 \right), & \mathrm{mod}(s,2) = 1 \end{cases} \tag{15}$$

Using Equation (13) and the backpropagation algorithm, the phase shift parameters $\theta_{s,t}$ can be optimized to determine the specific phase shift values for each phase shifter within the PMCC.

In this subsection, the training was performed on a single GPU, requiring approximately 3.5 h to converge with about 1×10^5 iterations. Following network training, we conducted classification simulations on the 10,000 test images, achieving a recognition accuracy of 99.05% with the confusion matrix shown in Figure 4d, thereby verifying the feasibility of this PMCC.

The PMCC architecture depicted in Figure 4a is specifically designed for the 0–9 handwritten digit classification task. For other task objectives, the architecture illustrated in Figure 1 can be adapted by modifying both the number of input/output ports ($2n$) and the corresponding count of OLC layers (n) to accommodate them. Notably, as n increases, the PMCC system constructed with the splitters and phase shifters designed in Sections 3 and 4 exhibits an approximately linear increase in insertion loss ($\sim$0.4n dB), a quadratic increase in training complexity, and a near-exponential rise in power consumption.

6. Stochastic Fabrication-Error Simulation Model

We have established a simulation-verified efficient PMCC achieving 99.05% accuracy under ideal conditions. We next turn to detailed analysis of potential error sources within this system and construct a stochastic error simulation framework to validate the PMCC's robustness.

The first device that affects the output of the PMCC is the power splitter. For the compact fully-symmetric power splitter designed using the GLINT algorithm, potential fabrication errors primarily originate from the loss of numerous small-scale isolated island/hole structures during the etching process. As demonstrated in our previous work, the presence of a high density of hole/island structures in the designed layout poses significant challenges to manufacturability. During the global search and local refinement steps of the GLINT algorithm, overlapping regions between individual circular search areas may lead to the formation of such small-scale isolated islands or holes.

To simulate fabrication errors in the power splitter, we introduced the following modifications to the design: removing all isolated islands and holes smaller than 40 nm, merging features separated by gaps narrower than 40 nm, and smoothing structural boundaries to emulate the worst-case fabricated device morphology. Ultimately, the resulting structure, shown in Figure 5b, is deemed to represent the power splitter geometry under conditions of maximum fabrication error. The electric field transmission matrix (transfer function) of this structure is defined as $\mathbf{B_E}$:

$$\mathbf{B_E} = \begin{bmatrix} -0.675 + 0.064i & -0.064 - 0.675i \\ -0.064 - 0.675i & -0.675 + 0.064i \end{bmatrix} \quad (16)$$

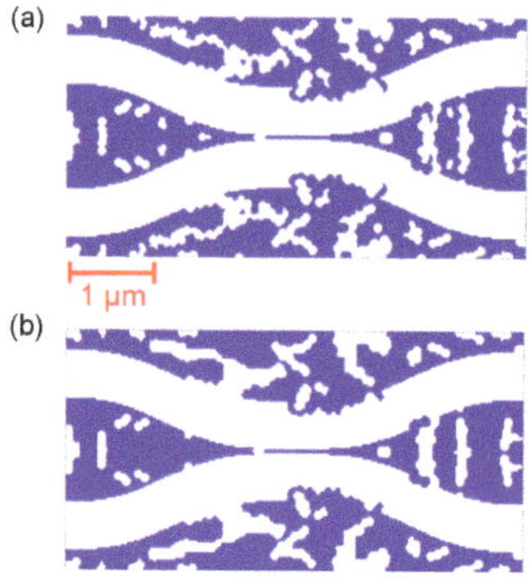

Figure 5. Schematic of fabrication error simulation. (**a**) Schematic of the designed compact power splitter; (**b**) schematic of the power splitter structure with simulated fabrication errors.

$\mathbf{B_E}$ represents the electric field transmission matrix under conditions of maximum fabrication error, where each element in $\mathbf{B_E}$ exhibits a significant deviation in both magnitude and phase compared to the original matrix $\mathbf{A_E}$. Here, we define the phase matrix of the electric field transmission matrix $\mathbf{\Gamma_E}$ as $\boldsymbol{\theta}_{\boldsymbol{\Gamma}\mathbf{E}}$:

$$\theta_{\Gamma E}(i,j) \ = arg(\Gamma_E(i,j)) \quad (17)$$

where $0 \le \theta_{\Gamma E}(i,j) < 2\pi$; $i, j = 1, 2$; $\Gamma = A, B$. Building on this, a stochastic electric field transmission matrix $\widetilde{\mathbf{A}}$ is generated by introducing Gaussian-distributed random variables ζ_{ij} and η_{ij}, $\zeta_{ij}, \eta_{ij} \sim N\left(0, \frac{1}{9}\right)$:

$$\widetilde{A}(i,j) = (|A_E(i,j)| + |\zeta_{ij}|(|B_E(i,j)| - |A_E(i,j)|))e^{i(\theta_{AE}(i,j)+|\eta_{ij}|(\theta_{BE}(i,j)-\theta_{AE}(i,j)))} \quad (18)$$

The magnitude and phase of each element in $\widetilde{\mathbf{A}}$ are defined as $\left|\widetilde{A}(i,j)\right|$ and $\theta_{\widetilde{A}}(i,j)$, respectively. In matrices $\mathbf{A_E}$ and $\mathbf{B_E}$, the squared modulus $|A_E(i,j)|^2$ and $|B_E(i,j)|^2$ represent the ideal transmission and the transmission under maximum fabrication error, respectively, while the phases $\theta_{AE}(i,j)$ and $\theta_{BE}(i,j)$ correspond to the ideal phase shift and the phase shift under maximum fabrication error of the electric field, respectively.

According to the 3σ rule of Gaussian distribution, the probabilities that $0 \le |\zeta_{ij}| < 1$ and $0 \le |\eta_{ij}| < 1$ are both 99.74%. Therefore, according to Equation (18), the probability that both $|A_E(i,j)| \le \left|\widetilde{A}(i,j)\right| < |B_E(i,j)|$ and $\theta_{AE}(i,j) \le \theta_{\widetilde{A}}(i,j) < \theta_{BE}(i,j)$ is 99.74%, thus effectively emulating the impact of stochastic fabrication errors on the electric field transmission matrix of the power splitter.

The second device that impacts the output of the PMCC is the phase shifter. Owing to its relatively regular structure, this compact phase shifter incorporates no challenging-to-fabricate features—such as holes, isolated islands, or narrow gaps—thus avoiding fabrication-induced loss of critical structures caused by complex geometries. The dominant source of error in this device stems from fabrication deviations in waveguide width, specifically over-etching or under-etching. Here, we set the maximum etching error of the phase shifter to $\pm$ 20 nm, corresponding to a phase deviation of $\pi/20$. Similarly, the designed phase θ is perturbed via a Gaussian random variable ξ, $\xi \sim N\left(0, \frac{1}{9}\right)$ to generate a stochastic phase $\tilde{\theta}$:

$$\tilde{\theta} = \theta + \frac{\pi}{20} \times \xi \tag{19}$$

The probability that $\theta - \pi/20 < \tilde{\theta} < \theta + \pi/20$ is 99.74%.

At this stage, we obtained the stochastic electric field transmission matrix $\tilde{\mathbf{A}}$ for the power splitters and the stochastic phase $\tilde{\theta}$ for the phase shifters. To validate the system robustness, 55 power splitter electric field transmission matrices $\mathbf{A_E}$ in the trained PMCC were replaced with $\tilde{\mathbf{A}}_u$ (u = 1, 2, ..., 55), and all phase shifter values $\theta_{s,t}$ were substituted with $\tilde{\theta}_{s,t}$ (s,t = 1, 2, ..., 10). After random assignment of 55 $\tilde{\mathbf{A}}_u$ matrices and 100 $\tilde{\theta}_{s,t}$ phase shifter values, a PMCC structure with random fabrication errors was obtained. We constructed 1000 PMCC structures with independent fabrication errors and evaluated the recognition accuracy for each PMCC using 10,000 test images, thereby obtaining the error-affected accuracy (absolute recognition accuracy) for 1000 individual PMCC with different fabrication errors. This procedure is equivalent to performing 1000 Monte Carlo trials.

Here, we use probability to represent the results of 1000 independent simulations of error-affected accuracy (Figure 6). The results show that the probability of the error-affected accuracy exceeds 90% is 0.689, and the probability that it exceeds 80% reaches 0.912. This indicates that, despite performance fluctuations induced by fabrication errors, the system maintains functional recognition capability with high probability. It is particularly noteworthy that even under the worst-case fabrication conditions—where the power splitter electric field transmission matrices approach $\mathbf{B_E}$ and each phase shifter exhibits the maximum etching error of $\pm$20 nm—the system still achieves recognition accuracy greater than 50%, demonstrating strong robustness of the PMCC against fabrication errors.

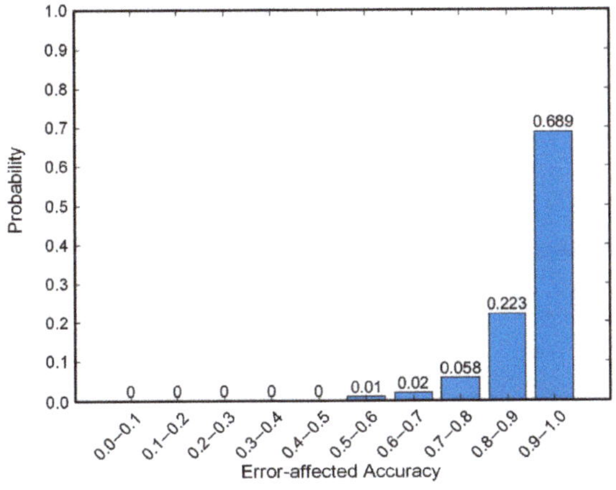

Figure 6. Probability distribution of error-affected accuracy from 1000 PMCC structures with random fabrication errors.

7. Conclusions

In summary, we have constructed an ultra-compact inverse-designed integrated PMCC on an SOI platform, which is capable of facilitating sophisticated functional matrix operations. In this work, the PMCC achieved a recognition accuracy of 99.05% on the 0–9 handwritten digit classification task. The proposed architecture comprises a networked cascade of power splitters designed with the GLINT algorithm and compact phase shifters, enabling matrix output modulation through multistage interference control, and achieves an integration density of 2.6×10^4 computational units per mm^2—far exceeding that of conventional multistage interferometric architectures such as MZI-based ONNs.

To validate the fabrication robustness of the PMCC, we developed a unique stochastic fabrication-error framework and performed 1000 PMCC structures with different fabrication errors by introducing random perturbations into the trained model, with their error-affected accuracy subsequently evaluated. Simulation results indicate that over 90% of the results maintain a recognition accuracy above 80%, and even under worst-case fabrication conditions where all fabrication errors are maximized, the accuracy remains above 50%, demonstrating the robustness of the PMCC.

Currently, the integration density of our PMCC is primarily constrained by the sizes of the power splitters and phase shifters. In the future, we will further optimize the GLINT algorithm to design even smaller and more manufacturable compact power splitters and phase shifters, thereby further increasing the integration density. We anticipate that with continued advances in etching and integration technologies, our PMCC framework will play an increasingly important role in the near future, particularly in edge AI applications such as real-time translation and autonomous driving.

Author Contributions: Conceptualization, M.X. and M.L.; methodology, M.X.; software, M.X.; validation, M.L., T.W., Y.Z., Y.S., J.Y. (Jie Yang), D.P., K.Z. and J.Y. (Jiahui Yao); formal analysis, M.L.; investigation, M.L.; resources, M.L.; data curation, M.L.; writing—original draft preparation, M.L.; writing—review and editing, M.X.; visualization, M.L.; supervision, M.X.; project administration, M.X.; funding acquisition, M.X. All authors have read and agreed to the published version of the manuscript.

Funding: This work was supported by the National Key R&D Program of China (Grant No. 2021YFC2201902) and National Natural Science Foundation of China (Grant No. 61975149).

Institutional Review Board Statement: Not applicable.

Informed Consent Statement: Not applicable.

Data Availability Statement: The authors will supply the relevant data in response to reasonable requests.

Conflicts of Interest: The authors declare no conflicts of interest.

References

1. Krizhevsky, A.; Sutskever, I.; Hinton, G.E. ImageNet classification with deep convolutional neural networks. *Commun. ACM* **2017**, *60*, 84–90. [CrossRef]
2. Voulodimos, A.; Doulamis, N.; Doulamis, A.; Protopapadakis, E. Deep learning for computer vision: A brief review. *Comput. Intell. Neurosci.* **2018**, *2018*, 7068349. [CrossRef]
3. Gu, J.; Neubig, G.; Cho, K.; Li, V. Learning to translate in real-time with neural machine translation. *Mach. Transl.* **2016**, *21*, 209–252.
4. Cui, Y.; Chen, R.; Chu, W.; Chen, L.; Tian, D.; Li, Y.; Cao, D. Deep learning for image and point cloud fusion in autonomous driving: A review. *IEEE Trans. Intell. Transp. Syst.* **2021**, *23*, 722–739. [CrossRef]
5. Naylor, M.; Runciman, C. The Reduceron: Widening the von Neumann bottleneck for graph reduction using an FPGA. In Proceedings of the Symposium on Implementation and Application of Functional Languages, Freiburg, Germany, 10–12 September 2007; pp. 129–146.

6. Shen, Y.; Harris, N.C.; Skirlo, S.; Prabhu, M.; Baehr-Jones, T.; Hochberg, M.; Sun, X.; Zhao, S.; Larochelle, H.; Englund, D. Deep learning with coherent nanophotonic circuits. *Nat. Photonics* **2017**, *11*, 441–446. [CrossRef]
7. Lin, X.; Rivenson, Y.; Yardimci, N.T.; Veli, M.; Luo, Y.; Jarrahi, M.; Ozcan, A. All-optical machine learning using diffractive deep neural networks. *Science* **2018**, *361*, 1004–1008. [CrossRef]
8. Feldmann, J.; Youngblood, N.; Karpov, M.; Gehring, H.; Li, X.; Stappers, M.; Le Gallo, M.; Fu, X.; Lukashchuk, A.; Raja, A.S. Parallel convolutional processing using an integrated photonic tensor core. *Nature* **2021**, *589*, 52–58. [CrossRef]
9. Fang, M.Y.-S.; Manipatruni, S.; Wierzynski, C.; Khosrowshahi, A.; DeWeese, M.R. Design of optical neural networks with component imprecisions. *Opt. Express* **2019**, *27*, 14009–14029. [CrossRef] [PubMed]
10. Zhu, H.; Zou, J.; Zhang, H.; Shi, Y.; Luo, S.; Wang, N.; Cai, H.; Wan, L.; Wang, B.; Jiang, X. Space-efficient optical computing with an integrated chip diffractive neural network. *Nat. Commun.* **2022**, *13*, 1044. [CrossRef]
11. Hughes, T.W.; Minkov, M.; Shi, Y.; Fan, S. Training of photonic neural networks through in situ backpropagation and gradient measurement. *Optica* **2018**, *5*, 864–871. [CrossRef]
12. Zarei, S.; Marzban, M.-r.; Khavasi, A. Integrated photonic neural network based on silicon metalines. *Opt. Express* **2020**, *28*, 36668–36684. [CrossRef]
13. Fu, T.; Zang, Y.; Huang, H.; Du, Z.; Hu, C.; Chen, M.; Yang, S.; Chen, H. On-chip photonic diffractive optical neural network based on a spatial domain electromagnetic propagation model. *Opt. Express* **2021**, *29*, 31924–31940. [CrossRef]
14. Fu, T.; Zang, Y.; Huang, Y.; Du, Z.; Huang, H.; Hu, C.; Chen, M.; Yang, S.; Chen, H. Photonic machine learning with on-chip diffractive optics. *Nat. Commun.* **2023**, *14*, 70. [CrossRef]
15. Liu, W.; Huang, Y.; Sun, R.; Fu, T.; Yang, S.; Chen, H. Ultra-compact multi-task processor based on in-memory optical computing. *Light Sci. Appl.* **2025**, *14*, 134. [CrossRef]
16. Meng, X.; Zhang, G.; Shi, N.; Li, G.; Azaña, J.; Capmany, J.; Yao, J.; Shen, Y.; Li, W.; Zhu, N. Compact optical convolution processing unit based on multimode interference. *Nat. Commun.* **2023**, *14*, 3000. [CrossRef]
17. Feldmann, J.; Youngblood, N.; Wright, C.D.; Bhaskaran, H.; Pernice, W.H. All-optical spiking neurosynaptic networks with self-learning capabilities. *Nature* **2019**, *569*, 208–214. [CrossRef] [PubMed]
18. Fu, T.; Zhang, J.; Sun, R.; Huang, Y.; Xu, W.; Yang, S.; Zhu, Z.; Chen, H. Optical neural networks: Progress and challenges. *Light Sci. Appl.* **2024**, *13*, 263. [CrossRef] [PubMed]
19. Molesky, S.; Lin, Z.; Piggott, A.Y.; Jin, W.; Vucković, J.; Rodriguez, A.W. Inverse design in nanophotonics. *Nat. Photonics* **2018**, *12*, 659–670. [CrossRef]
20. Lucas, E.; Yu, S.-P.; Briles, T.C.; Carlson, D.R.; Papp, S.B. Tailoring microcombs with inverse-designed, meta-dispersion microresonators. *Nat. Photonics* **2023**, *17*, 943–950. [CrossRef]
21. Yang, K.Y.; Skarda, J.; Cotrufo, M.; Dutt, A.; Ahn, G.H.; Sawaby, M.; Vercruysse, D.; Arbabian, A.; Fan, S.; Alù, A. Inverse-designed non-reciprocal pulse router for chip-based LiDAR. *Nat. Photonics* **2020**, *14*, 369–374. [CrossRef]
22. Su, L.; Piggott, A.Y.; Sapra, N.V.; Petykiewicz, J.; Vuckovic, J. Inverse design and demonstration of a compact on-chip narrowband three-channel wavelength demultiplexer. *ACS Photonics* **2018**, *5*, 301–305. [CrossRef]
23. Piggott, A.Y.; Lu, J.; Lagoudakis, K.G.; Petykiewicz, J.; Babinec, T.M.; Vučković, J. Inverse design and demonstration of a compact and broadband on-chip wavelength demultiplexer. *Nat. Photonics* **2015**, *9*, 374–377. [CrossRef]
24. Shen, B.; Wang, P.; Polson, R.; Menon, R. An integrated-nanophotonics polarization beamsplitter with $2.4 \times 2.4\ \mu m^2$ footprint. *Nat. Photonics* **2015**, *9*, 378–382. [CrossRef]
25. Shen, B.; Wang, P.; Polson, R.; Menon, R. Integrated metamaterials for efficient and compact free-space-to-waveguide coupling. *Opt. Express* **2014**, *22*, 27175–27182. [CrossRef]
26. Lu, J.; Vučković, J. Nanophotonic computational design. *Opt. Express* **2013**, *21*, 13351–13367. [CrossRef]
27. Li, M.; Wang, T.; Zhang, Y.; Shen, Y.; Yang, J.; Zhang, K.; Pan, D.; Xin, M. Global–Local Cooperative Optimization in Photonic Inverse Design Algorithms. *Photonics* **2025**, *12*, 725. [CrossRef]
28. Du, Z.; Liao, K.; Dai, T.; Wang, Y.; Gao, J.; Huang, H.; Qi, H.; Li, Y.; Wang, X.; Su, X. Ultracompact and multifunctional integrated photonic platform. *Sci. Adv.* **2024**, *10*, eadm7569. [CrossRef]

photonics

MDPI

Article

Phase-Matched Design for Efficient Entangled Photon Pair Generation in 3R-MoS_2 Waveguides

Shicheng Yu [1,2], Xiaojie Zhang [1,2], Xia Lei [1,2,*] and Liang Zhai [1,2,*]

1 Institute of Fundamental and Frontier Sciences, University of Electronic Science and Technology of China, Chengdu 610054, China
2 Tianfu Jiangxi Laboratory, Chengdu 641419, China
* Correspondence: xialei.june@outlook.com (X.L.); liang.zhai@uestc.edu.cn (L.Z.)

Abstract

Entangled photons are essential for photonic quantum technologies. Their generation typically relies on spontaneous parametric down-conversion, but conventional nonlinear crystals are bulky and hard to integrate on chips. Rhombohedral-stacked MoS_2 combines a high refractive index, large second-order nonlinearity, and flexibility for heterogeneous integration, making it a promising platform for integrated quantum photonics. However, the typical thin-film form of 3R-MoS_2 restricts the effective nonlinear interaction length, limiting entanglement generation efficiency in practical devices. To overcome this, phase-matching strategies in integrated waveguides are required but have so far remained undeveloped. Here, we introduce a waveguide-integrated 3R-MoS_2 platform with periodic grooves to achieve quasi-phase matching, enhancing down-conversion efficiency. Leveraging $\chi^{(2)}$ tensor symmetries and orthogonal waveguide modes, the design efficiently generates entangled photons, providing a compact, scalable route toward 2D-material-based integrated quantum photonic circuits.

Keywords: spontaneous parametric down-conversion; van-der waals materials; phase matching; entangled photons generation; 3R-MoS_2; integrated photonic waveguides

1. Introduction

Entangled photons are fundamental resources for quantum technologies, underpinning applications in communication, computation, and sensing. They can be generated either from quantum emitters, such as atoms [1] and quantum dots [2–4], or through nonlinear optical processes in bulk materials or on integrated platforms [5,6]. Among these approaches, spontaneous parametric down-conversion (SPDC), a second-order nonlinear process, remains the most widely used method for producing polarisation-entangled photons. Conventional bulk crystals such as BBO and KTP have been extensively employed [7,8], but their large dimensions and limited compatibility with photonic integration hinder the development of scalable and compact quantum circuits [9,10].

Rhombohedral-stacked two-dimensional (2D) materials, exemplified by 3R-MoS_2, have recently emerged as a promising new class of nonlinear media for integrated quantum photonics [11–15]. Unlike conventional bulk crystals, they combine a high refractive index for strong light confinement with a large second-order nonlinearity ($\chi^{(2)}\sim$100–800 pm/V) and intrinsic compatibility with chip-scale integration [15–18]. Wafer-scale growth of homogeneous rhombohedral films has now been demonstrated [19–21], strengthening their prospects for scalable photonic devices. Entangled photon pairs have already been realised, with tunability between different Bell states enabled by in-plane crystal symmetries [22–24]. Nevertheless, the limited thickness of 3R-MoS_2 (typically a few hundred nanometres) constrains the effective nonlinear interaction length. While such thin films do not generally require phase matching, the short interaction length still leads to modest conversion efficiency, albeit substantially higher than in monolayers. To achieve further enhancement, extending the interaction length in waveguide or cavity geometries will require effective phase-matching strategies [25–27]. Such phase-matching strategies have so far remained undeveloped in integrated 3R-MoS_2 waveguides.

Here, we introduce a waveguide-integrated 3R-MoS_2 platform that employs periodic grooves to realise quasi-phase matching (QPM). This approach modulates the effective refractive index along the propagation direction, compensating phase mismatch and markedly improving down-conversion efficiency. By leveraging the $\chi^{(2)}$ tensor symmetries of 3R-MoS_2, the design utilises orthogonal electric field components in different waveguide modes for entangled-photon generation. Our design offers a practical pathway toward compact, high-performance entangled-photon sources in 2D materials, advancing their integration into scalable quantum photonic circuits.

2. Design Strategy

We first focus on the principle of parametric down-conversion in 3R-MoS_2 and outline our design strategy. In rhombohedral MoS_2, the S–Mo–S monolayers are stacked in a translationally shifted sequence (ABC stacking, see Figure 1a), without the 180° rotations that appear in the 2H phase. This stacking breaks inversion symmetry along the vertical (out-of-plane, labelled y, Figure 1b) direction even for multilayer or thick 3R-MoS_2, allowing a non-zero second-order nonlinear susceptibility, $\chi^{(2)}$, to persist with increasing thickness. The crystal belongs to the non-centrosymmetric point group D_{3h}. In this symmetry, the dominant non-vanishing $\chi^{(2)}$ tensor components [18,22] are $\chi^{(2)}_{zxx} = \chi^{(2)}_{xzx} = \chi^{(2)}_{xxz} = -\chi^{(2)}_{zzz}$, which couple primarily to the in-plane electric field components (E_x, E_z), whereas contributions involving the out-of-plane field (E_y) are typically much weaker.

The symmetry of the $\chi^{(2)}$ tensor underpins the generation of polarisation-entangled photon pairs. In the SPDC process, an x-polarised pump photon can produce either x_s–z_i photon pairs via $\chi^{(2)}_{xzx}$ or z_s–x_i pairs via $\chi^{(2)}_{xxz}$ (where the subscripts s and i denote the signal and idler photons), with the same principle applying to a z-polarised pump. This enables a coherent superposition, yielding the entangled state $|\Psi\rangle = (|x\rangle\,|z\rangle + |z\rangle\,|x\rangle)/\sqrt{2}$ under an x-polarised pump. These tensor properties have important implications for integrated device design, as the polarisation-dependent coupling must be carefully controlled. In particular, the down-converted photons must be coupled to orthogonal electric field components—if one photon couples to E_x, the other must couple to E_z—necessitating separate mode design.

Efficient nonlinear energy conversion further requires satisfying the phase-matching condition for propagation in waveguides, $\Delta k = 0$ with $\mathbf{k}_p = \mathbf{k}_s + \mathbf{k}_i$. For practical applications in quantum communication, we target entangled-photon generation in the telecom band around 1560 nm. At this wavelength, the intrinsic material absorption is negligible [12]. However, achieving phase matching in 3R-MoS_2 is challenging due to its strong birefringence and material dispersion. The optical axis lies along the y direction, with refractive indices [12] of $n_o \approx 4.7$ and $n_e \approx 2.7$ at 780 nm, while at 1560 nm n_o decreases to $\approx$4.1 and n_e remains nearly constant. These material properties preclude straightforward birefringent phase matching in the waveguide, necessitating alternative strategies to achieve high-efficiency entangled photon-pair generation.

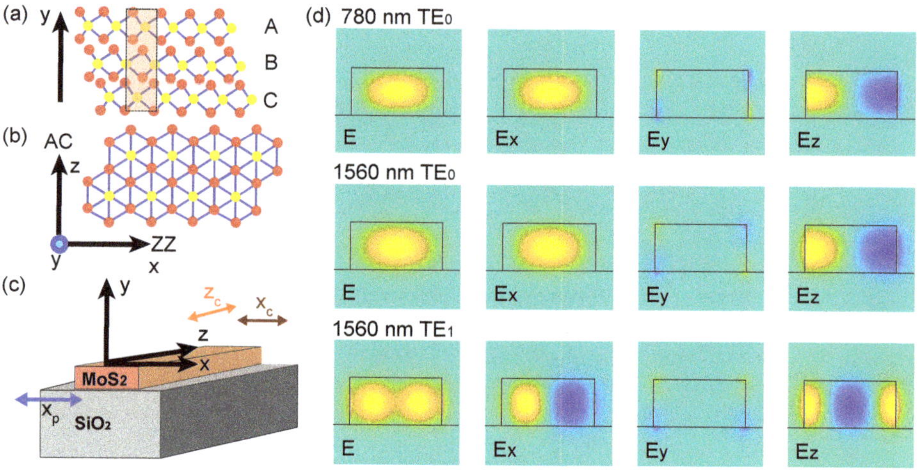

Figure 1. The strip waveguide design. (**a**,**b**) Atomic structure of 3R-MoS_2. A, B, and C indicate the lateral positions of successive layers, forming an ABC–ABC stacking sequence distinct from the AB–AB stacking of 2H-MoS_2. The y-axis is defined along the vertical stacking direction, while the x–z plane coincides with the MoS_2 surface, aligned with the armchair (AC) and zigzag (ZZ) crystal directions. The dominant non-vanishing $\chi^{(2)}$ tensor components lie within the x–z plane. (**c**) Schematic of the 3R-MoS_2 thin film in a strip-waveguide geometry. The z-axis represents the propagation direction, and the x–y plane corresponds to the cross-sectional view. (**d**) Electric field distributions and their projections along the x, y, and z axes for the TE_0 (780 nm), TE_0 (1560 nm), and TE_1 (1560 nm) modes, respectively. The mode profiles are calculated using a finite-difference eigenmode (FDE) solver in Lumerical MODE. Yellow, cyan, and blue indicate positive, zero, and negative field amplitudes, respectively.

3. Strip Waveguide Design

We first consider a simple rectangular (strip) waveguide of 3R-MoS_2 that is transferred on a SiO_2 substrate. The effective refractive index is modified by the modal distribution of the waveguide, which in turn affects the wavevectors.

The waveguide cross-section is aligned with the material's x–y plane, with the z-axis defined as the propagation direction (Figure 1c). Note that the coordinate system follows that in Figure 1a,b, where the y-axis corresponds to the ABC stacking direction (vertical thickness). In this configuration, the SPDC process follows the notation introduced earlier: $|x\rangle_p \rightarrow |x\rangle_s |z\rangle_i + |z\rangle_s |x\rangle_i$, where the pump and entangled photons are at 780 nm and 1560 nm (degenerate), respectively. At a given position z^j, the SPDC efficiency is determined by the transverse overlap of the electric fields, $\eta(z^j) \propto |\int_S \chi^{(2)}_{xzx} E^p_x(z^j)^* E^c_x(z^j) E^c_z(z^j) dS|^2$, where the superscripts p, c denote the pump and down-converted photons, respectively, and the integral is taken over the waveguide cross-section. The design therefore aims to select modes at 1560 nm with strong E_x and E_z components to maximise the nonlinear overlap while keeping the entangled photons in separable orthogonal modes.

The electric field distribution of the fundamental TE_0 mode at the pump wavelength of 780 nm was analysed within this waveguide (Figure 1d, first row). The results show that the electric field is dominated by the x-component (E_x), while the z-component is much weaker and exhibits odd symmetry along the x-direction, and the y-component is negligible. A similar distribution is obtained for the fundamental TE_0 mode at the conversion wavelength of 1560 nm, except that the confinement at this longer wavelength is slightly weaker, leading to a broader spatial extent of the mode profile and slightly reduced field intensity in the waveguide region (Figure 1d, second row).

The third row of Figure 1d displays the electric field distribution of a higher order mode, TE_{10}. Since the waveguide width is assumed to be larger than its height, the mode exhibits a zero crossing (node) along the x-direction (the width). We thus refer to it as TE_1 later, as TE_{01} is less likely to occur. Here, the x-component remains the strongest but has odd symmetry about the centre of the waveguide. This results in cancellation in the cross-sectional SPDC integral, since $E_x^p(TE_0)^* E_x^c(TE_0) E_x^c(TE_1)$ is an odd function, giving an integral of zero. In contrast, the z-component, which couples to E_x^p via the $\chi^{(2)}_{xxz}$ tensor, exhibits even symmetry across x, allowing a non-zero contribution. This ensures that the down-converted photons are effectively separable into E_x (TE_0) and E_z (TE_1) modes, with negligible unwanted coupling, since E_z of TE_0 and E_x of TE_1 are both odd.

We proceed to analyse the modal refractive indices, n_{eff}, of these modes. Both the waveguide width and height were scanned, and the corresponding modal indices were extracted [see Figure 2a–c, with contours indicated by white dashed lines, separated by steps of 0.25]. Across all modes, n_{eff} increases with increasing waveguide width and height [see Figure 2d,e], as expected, since larger waveguide dimensions confine the optical mode more effectively. It should be noted that, because the E_y components of the modes are very small, the material's extraordinary refractive index has minimal impact on n_{eff}, although it is still included in the calculations.

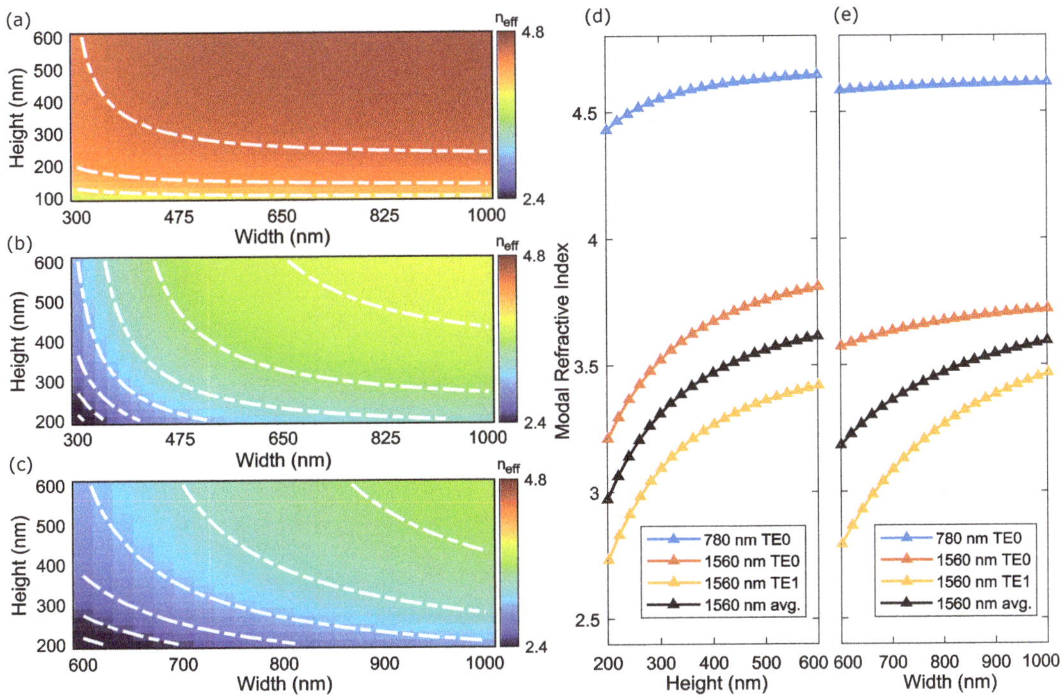

Figure 2. Modal refractive index of the strip waveguide. (**a–c**) Extracted modal refractive indices n_{eff} of the ridge waveguide: TE_0 mode at the pump wavelength (**a**), TE_0 mode at the conversion wavelength (**b**), and TE_1 mode at the conversion wavelength (**c**) for varying waveguide parameters, with white dashed lines indicating n_{eff} contours. (**d,e**) Extracted modal refractive indices as functions of the waveguide height (**d**) and width (**e**), respectively. In (**d**), the waveguide width is fixed at 800 nm, whereas in (**e**), the height is fixed at 400 nm. The modal indices are obtained using a finite-element method (FEM) solver in COMSOL Multiphysics 5.4.

The strongly confined TE_0 mode at 780 nm exhibits a relatively high refractive index (denoted as $n^p_{TE_0}$), ranging from approximately 3.8 for very shallow waveguides to around 4.6 for typical waveguide dimensions (e.g., width∼800 nm, height∼400 nm). In contrast, n_{eff} is lower for the 1560 nm modes, with TE_0 values (denoted as $n^c_{TE_0}$) ranging from about 2.4 (width 300 nm, height 200 nm) to 3.7 (width 800 nm, height 400 nm), and TE_1 values (denoted as $n^c_{TE_1}$) between 2.15 (width 600 nm, height 200 nm) and 3.6 (width 1000 nm, height 600 nm).

For the same waveguide height and width, the n_{eff} values of the three modes are compared as functions of waveguide width and height, respectively, while keeping the other parameter fixed. To satisfy the phase-matching condition, $\Delta k = 2\pi(n^p_{TE_0}/\lambda^p - n^c_{TE_0}/\lambda^c - n^c_{TE_1}/\lambda^c)$, the modal indices must strictly obey $n^c_{avg} \equiv (n^c_{TE_0} + n^c_{TE_1})/2 = n^p_{TE_0}$. In Figure 2d,e, $n^p_{TE_0}$ is plotted in blue and n^c_{avg} in black. Clearly, no crossing occurs, indicating that the modal phase-matching condition cannot be satisfied. This is primarily because the strongly confined TE_0 pump mode has a refractive index much higher than the 1560 nm modes, as evidenced by the colour differences in Figure 2a–c. Although the pump n_{eff} could in principle be lowered by employing higher-order modes such as TE_2 or TE_4, these modes considerably complicate the design, since the Gaussian input beam cannot efficiently excite them without a dedicated modal converter [28,29].

4. Quasi-Phase Matching Waveguide Design

The mismatch between the modal refractive indices, albeit geometric optimisation, necessitates the adoption of alternative approaches to achieve phase matching. One well-established technique is quasi-phase matching [5,30–32], which has been widely implemented in lithium niobate and KTP, where the index mismatch is compensated by periodic inversion of the nonlinear domain. Similar concepts have been applied to rhombohedrally stacked MoS_2, where the sign of $\chi^{(2)}$ is reversed through rotational stacking of 3R-MoS_2 layers along the vertical direction. However, this approach demands precise control of rotation angle and layer thickness, and so far has only extended the effective interaction length to the micrometre scale [26,27].

In general, the principle of quasi-phase matching can be understood from a Fourier perspective. For SPDC, the efficiency of generating photon pairs is determined by the longitudinal accumulation of the mode overlap factor $\Gamma(z^j) = \int_S \chi^{(2)}_{xzx} E^p_x(z^j)^* E^c_x(z^j) E^c_z(z^j) dS$, which quantifies the transverse overlap between the pump and the target signal and idler modes. The accumulated conversion amplitude along the propagation direction for a total length L is given by $A_c \propto \int_0^L \Gamma(z^j) e^{i\Delta k z} dz$. When $\Delta k \neq 0$, the exponential causes partial cancellation and limits the growth. Quasi-phase matching compensates for the phase mismatch by ensuring that the effective mode overlap along z contains a periodic modulation whose Fourier component at Δk provides constructive addition of the contributions from different positions.

In conventional domain-inverted QPM, $\chi^{(2)}(z^j)$ flips between $\pm\chi^{(2)}$, producing a square-wave modulation with a strong Fourier peak at $2\pi/\Lambda \approx \Delta k$. More generally, literal sign inversion is not required: any longitudinal modulation of the effective mode overlap that produces Fourier weight at Δk partially compensates the phase mismatch and allows gradual accumulation of photon pairs. In our design, the waveguide is grooved (schematic in Figure 3c)—its width is modulated with period Λ, rather than flipping $\chi^{(2)}$, where $\Lambda = 2\pi/\Delta k$. This periodic grooving alters the mode overlap factor $\Gamma(z^j)$, producing a Fourier component at $k_{QPM} = 2\pi/\Lambda$ that compensates the phase mismatch, $k_p = k_s + k_i + k_{QPM}$ [Figure 3a,b].

As a starting point, we choose the dimensions of the unaltered waveguide to be 800 nm in width and 400 nm in height. These values are based on considerations of small phase mismatch and practical feasibility in fabrication. We first verify the electric field distribution within the periodically grooved waveguides, selecting a groove ratio of 15% (which affects the efficiency, as will be discussed later). The distributions of E_x^p (TE_0), E_x^c (TE_0), and E_z^c (TE_1) are shown for both the unaltered and grooved regions of the waveguide in Figure 3d. Despite the differences between the two, they still resemble the corresponding distributions in the strip waveguide shown in Figure 1d. This indicates that the polarisation entanglement considerations discussed previously in Section 3 remain valid.

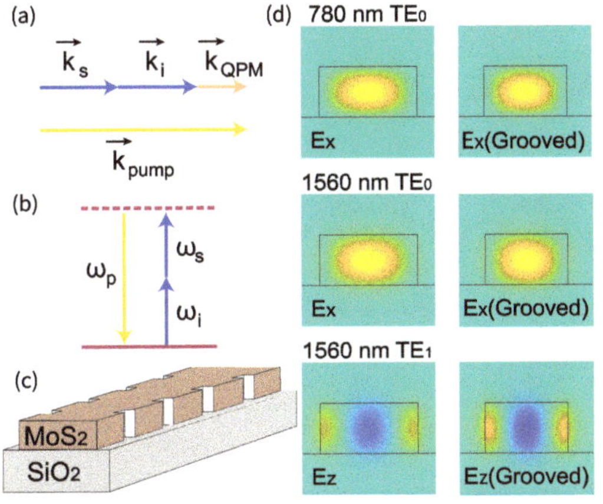

Figure 3. The waveguide design with periodic grooving. (**a**) Periodic modulation of the waveguide width alters the overlap of the transverse electric fields, providing a Fourier component $\vec{k}_{QPM}$ that compensates for the phase mismatch. (**b**) The energy of the conversion must still be conserved. (**c**) Schematic of the 3R-MoS_2 thin film in a periodically grooved waveguide geometry on a SiO_2 substrate. (**d**) Electric field distributions along the x, x, and z axes for the TE_0 (780 nm), TE_0 (1560 nm), and TE_1 (1560 nm) modes, respectively. The mode profiles are calculated using an FDE solver in Lumerical MODE. Yellow, cyan, and blue indicate positive, zero, and negative field amplitudes, respectively.

We now examine the effect of the groove ratio and explain our considerations. Here, the groove ratio denotes the ratio of the waveguide width in the grooved region to that in the ungrooved region. The modal refractive indices are extracted for varying groove ratios, as shown in Figure 4a. Increasing the ratio reduces the waveguide width in the grooved region, effectively lowering n_{eff}, with the higher-order TE_1 mode decreasing more rapidly than TE_0. This trend is fully consistent with Figure 2e.

The modification of n_{eff} due to the introduction of grooving necessitates consideration of the local phase mismatch in each section. We therefore define an effective phase mismatch as $\Delta k_{eff} = d\Delta k + (1-d)\Delta k'$, where Δk and $\Delta k'$ correspond to the ungrooved and grooved regions and d is the duty cycle (set to 50% in our analysis). The QPM period Λ is then chosen such that $\Lambda = 2\pi/\Delta k_{eff}$, ensuring constructive buildup of the nonlinear signal over multiple modulation cycles.

We numerically modelled the down-conversion process in a grooved waveguide over a propagation length of 100 μm. Without grooving, the conversion amplitude oscillates with a characteristic length scale given by the coherence length $L_c = \pi/\Delta k \approx 0.34$ μm, and thus does not accumulate with increasing propagation length (Figure 4d). Introducing grooves into the waveguide adds a modulation component at the corresponding Fourier peak, effectively compensating for the phase mismatch and enabling the down-conversion amplitude to grow cumulatively.

In Figure 4c, we model the evolution of the down-conversion amplitude for different groove ratios. For each ratio, the modulation period Λ is adjusted to match the corresponding phase mismatch Δk, since varying the groove ratio modifies the waveguide confinement and thereby alters the modal refractive indices. The trend is clear: for groove ratios between 0% and 15%, a larger ratio enhances the modulation component at the relevant Fourier peak, leading to faster growth of the amplitude. At a groove ratio of 15%, the down-conversion amplitude is approximately 160 times larger than that without grooving, providing a clear verification of our approach. The conversion amplitude after 100 μm of propagation is shown as a function of groove ratio in Figure 4b, revealing a monotonic increase up to a 0.15 groove ratio. Further increasing the groove ratio could potentially boost the conversion amplitude, but at the cost of weaker mode confinement (Figure 4a) and higher Bloch scattering losses [29].

Note that our analyses above focus on the creation of $|\Psi\rangle = (|x\rangle|z\rangle + |z\rangle|x\rangle)/\sqrt{2}$ under x-polarised pump. Another Bell state, $|\Phi\rangle = (|x\rangle|x\rangle - |z\rangle|z\rangle)/\sqrt{2}$, can also be generated by employing z-polarised pump via the $\chi^{(2)}_{zxx}$ or $\chi^{(2)}_{zzz}$ tensors. A similar waveguide design, with the propagation direction oriented along the x-axis, could in principle provide a QPM condition. Nevertheless, simultaneously retaining the tunability between Bell states while achieving phase matching remains an ongoing research challenge.

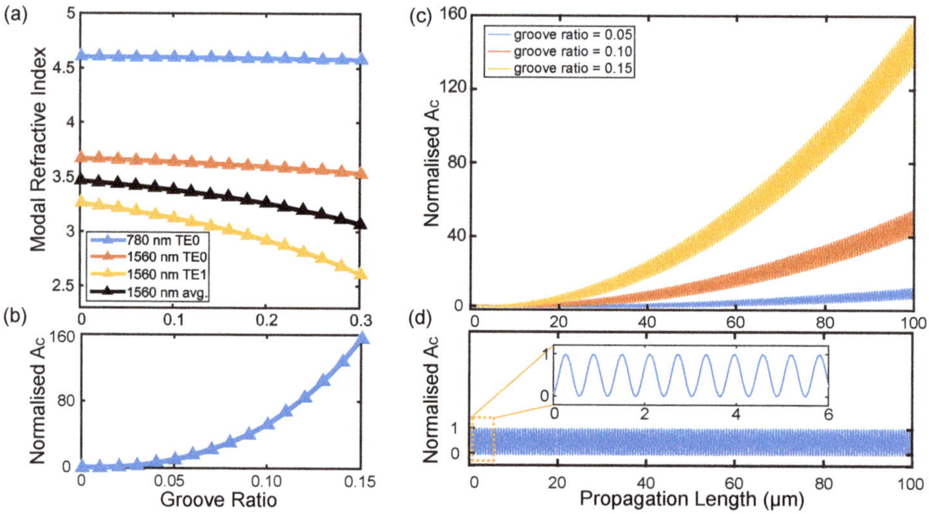

Figure 4. Efficient down-conversion through periodic waveguide grooving. (**a**) Extracted modal refractive indices of the relevant optical modes in the grooved waveguide. (**b**) Conversion amplitude (denoted as A_c) after 100 μm propagation for different groove ratios. For each groove ratio, the groove period is readjusted to match the corresponding phase mismatch. (**c**) Conversion amplitude as a function of propagation length, showing oscillatory growth for different groove ratios. The amplitudes are normalised to the oscillation amplitude of the ungrooved case. (**d**) Conversion amplitude in a strip waveguide without grooving, where the amplitude oscillates but does not accumulate with propagation. The inset shows a zoom-in of the short-propagation regime (0–6 μm), clearly resolving the oscillations. The oscillation period is $2L_c \approx 0.68$ μm. For (**a**–**d**), modal indices and electric field distributions are first calculated using an FDE solver in Lumerical MODE, and the conversion amplitudes are evaluated based on the overlap of these field distributions. Losses due to Bloch scattering and material absorption are neglected because of the short propagation distance and relatively small groove ratios.

5. Discussion

To summarise, we design a phase-matched approach for polarisation-entangled photon generation in integrated 3R-MoS_2 waveguides. The polarisation entanglement is preserved by separating the signal and idler photons into the E_x and E_z components of the TE_0 and TE_1 modes, respectively. Phase matching is achieved through periodic modulation of the waveguide width, with a period precisely designed to compensate for the intrinsic phase mismatch. This approach provides a practical strategy to enhance the nonlinear conversion efficiency of 3R-MoS_2 on an integrated platform, addressing a key limitation of 2D-material-based nonlinear photonics. Beyond this proof-of-concept, our design can be combined with advanced nanofabrication techniques and optimised waveguide engineering—such as mode-shape tailoring, dispersion engineering, or hybrid integration with low-loss photonic platforms [33,34]—to realise high-performance, low-loss on-chip entangled-photon sources. Furthermore, the flexibility and tunability of 3R-MoS_2 open opportunities to extend this approach to other 2D materials or heterostructures [11,20,35], potentially covering a broad range of wavelengths and enabling multi-band or multiplexed quantum light sources. Importantly, the strategy is inherently compatible with material transfer and hetero-integration [36,37], allowing combination with mature integrated photonics platforms, single-photon detectors [38], or deterministic quantum emitters [39–41], further advancing scalable and versatile quantum photonic technologies.

Author Contributions: Conceptualisation, S.Y., X.L. and L.Z.; methodology, S.Y., X.L. and L.Z.; software, S.Y. and L.Z.; validation, S.Y., X.L. and L.Z.; formal analysis, S.Y., X.Z. and L.Z.; investigation, S.Y., X.Z. and L.Z.; resources, L.Z.; writing—original draft preparation, S.Y., X.L. and L.Z.; writing—review and editing, X.L. and L.Z.; visualisation, S.Y., X.Z. and L.Z.; supervision, L.Z.; project administration, L.Z.; funding acquisition, L.Z. All authors have read and agreed to the published version of the manuscript.

Funding: This work was supported by the Sichuan Science and Technology Programme under Grant Nos. 2024ZYD0050 and 2024YFHZ0369.

Data Availability Statement: The raw data that support the findings of this study are available at https://doi.org/10.5281/zenodo.17552810 and from the corresponding author upon reasonable request.

Acknowledgments: We thank Qing Zhang and Yuqi Wang for their valuable discussions.

Conflicts of Interest: The authors declare that they have no competing interests.

References

1. Thomas, P.; Ruscio, L.; Morin, O.; Rempe, G. Fusion of deterministically generated photonic graph states. *Nature* **2024**, *629*, 567–572. [CrossRef]
2. Yu, S.; Zhang, X.; Lei, X.; Zhai, L. A Concise Primer on Solid-State Quantum Emitters. *IEEE J. Sel. Top. Quantum Electron.* **2025**, *31*, 6700616. [CrossRef]
3. Liu, S.; Wang, Y.; Saleem, Y.; Li, X.; Liu, H.; Yang, C.A.; Yang, J.; Ni, H.; Niu, Z.; Meng, Y.; et al. Quantum correlations of spontaneous two-photon emission from a quantum dot. *Nature* **2025**, *643*, 1234–1239. [CrossRef]
4. Laneve, A.; Rota, M.B.; Basso Basset, F.; Beccaceci, M.; Villari, V.; Oberleitner, T.; Reum, Y.; Krieger, T.M.; Buchinger, Q.; Prasad, R.; et al. Wavevector-resolved polarization entanglement from radiative cascades. *Nat. Commun.* **2025**, *16*, 6209. [CrossRef] [PubMed]
5. Lin, J.; Yao, N.; Hao, Z.; Zhang, J.; Mao, W.; Wang, M.; Chu, W.; Wu, R.; Fang, Z.; Qiao, L.; et al. Broadband quasi-phase-matched harmonic generation in an on-chip monocrystalline lithium niobate microdisk resonator. *Phys. Rev. Lett.* **2019**, *122*, 173903. [CrossRef] [PubMed]
6. Larsen, M.V.; Collins, M.J.; Broome, M.A.; Strain, M.J.; Tame, M.S.; Almeida, M.P.; Reid, M.D.; White, A.G. Deterministic generation of a two-dimensional cluster state. *Science* **2019**, *365*, 369–372. [CrossRef] [PubMed]
7. Lu, C.Y.; Yang, T.; Pan, J.W. Experimental Multiparticle Entanglement Swapping for Quantum Networking. *Phys. Rev. Lett.* **2009**, *103*, 020501. [CrossRef]

8. Villar, A.; Lohrmann, A.; Ling, A. Experimental entangled photon pair generation using crystals with parallel optical axes. *Opt. Express* **2018**, *26*, 12396. [CrossRef]
9. Elshaari, A.W.; Pernice, W.; Srinivasan, K.; Benson, O.; Zwiller, V. Hybrid integrated quantum photonic circuits. *Nat. Photonics* **2020**, *14*, 285–298. [CrossRef]
10. Zhou, X.; Zhai, L.; Liu, J. Epitaxial quantum dots: A semiconductor launchpad for photonic quantum technologies. *Photonics Insights* **2023**, *1*, R07. [CrossRef]
11. Autere, A.; Jussila, H.; Dai, Y.; Wang, Y.; Lipsanen, H.; Sun, Z. Nonlinear optics with two-dimensional layered materials. *Adv. Mater.* **2018**, *30*, 1705963. [CrossRef]
12. Xu, X.; Trovatello, C.; Mooshammer, F.; Shao, Y.; Zhang, S.; Yao, K.; Basov, D.N.; Cerullo, G.; Schuck, P.J. Towards compact phase-matched and waveguided nonlinear optics in atomically layered semiconductors. *Nat. Photonics* **2022**, *16*, 698–706. [CrossRef]
13. Tang, J.; Gao, M.; Zhang, Q.; Yang, Y. Giant second-harmonic generation in rhombohedral transition metal dichalcogenide metasurfaces. *Opt. Commun.* **2025**, *584*, 131825. [CrossRef]
14. Xu, D.; Peng, Z.H.; Trovatello, C.; Cheng, S.W.; Xu, X.; Sternbach, A.; Basov, D.N.; Schuck, P.J.; Delor, M. Spatiotemporal imaging of nonlinear optics in van der Waals waveguides. *Nat. Nanotechnol.* **2025**, *20*, 374–380. [CrossRef] [PubMed]
15. Tang, Y.; Qin, H.; de Ceglia, D.; Yang, W.; Shameli, M.A.; Nauman, M.; Morales, R.C.; Yan, J.; Wang, C.; Qiu, S.; et al. Giant Second Harmonic Generation from 3R-MoS_2 Metasurfaces. *arXiv* **2025**, arXiv:2503.20161.
16. Trolle, M.L.; Tsao, Y.C.; Pedersen, K.; Pedersen, T.G. Observation of excitonic resonances in the second harmonic spectrum of MoS_2. *Phys. Rev. B* **2015**, *92*, 161409. [CrossRef]
17. Zograf, G.; Küçüköz, B.; Polyakov, A.Y.; Yankovich, A.B.; Ranjan, A.; Bancerek, M.; Agrawal, A.V.; Olsson, E.; Wieczorek, W.; Antosiewicz, T.J.; et al. Ultrathin 3R-MoS_2 metasurfaces with atomically precise edges for efficient nonlinear nanophotonics. *Commun. Phys.* **2025**, *8*, 271. [CrossRef]
18. Seidt, L.; Weber, T.; Seredin, A.A.; Possmayer, T.; Savelev, R.; Petrov, M.I.; Maier, S.A.; Tittl, A.; de S. Menezes, L.; Sortino, L. Ultrafast all-optical switching in nonlinear 3R-MoS_2 van der Waals metasurfaces. *npj Nanophotonics* **2025**, *2*, 37. [CrossRef]
19. Qin, B.; Ma, C.; Guo, Q.; Li, X.; Wei, W.; Ma, C.; Wang, Q.; Liu, F.; Zhao, M.; Xue, G.; et al. Interfacial epitaxy of multilayer rhombohedral transition-metal dichalcogenide single crystals. *Science* **2024**, *385*, 99–104. [CrossRef]
20. Wang, L.; Qi, J.; Wei, W.; Wu, M.; Zhang, Z.; Li, X.; Sun, H.; Guo, Q.; Cao, M.; Wang, Q.; et al. Bevel-edge epitaxy of ferroelectric rhombohedral boron nitride single crystal. *Nature* **2024**, *629*, 74–79. [CrossRef]
21. Liu, L.; Li, T.; Gong, X.; Wen, H.; Zhou, L.; Feng, M.; Zhang, H.; Zou, N.; Wu, S.; Li, Y.; et al. Homoepitaxial growth of large-area rhombohedral-stacked MoS_2. *Nat. Mater.* **2025**, *24*, 1195–1202. [CrossRef]
22. Weissflog, M.A.; Fedotova, A.; Tang, Y.; Santos, E.A.; Laudert, B.; Shinde, S.; Abtahi, F.; Afsharnia, M.; Pérez Pérez, I.; Ritter, S.; et al. A tunable transition metal dichalcogenide entangled photon-pair source. *Nat. Commun.* **2024**, *15*, 7600. [CrossRef]
23. Feng, J.; Wu, Y.; Duan, R.; Wang, J.; Chen, W.; Qin, J.; Liu, Z.; Guo, G.C.; Ren, X.F.; Qiu, C.W. Polarization-entangled photon-pair source with van der Waals 3R-WS_2 crystal. *eLight* **2024**, *4*, 16. [CrossRef]
24. Lyu, X.; Kallioniemi, L.; Hong, H.; Qu, R.; Zhang, Y.; Zúñiga-Perez, J.; Liu, K.; Gao, W. A tunable entangled photon-pair source based on a Van der Waals insulator. *Nat. Commun.* **2025**, *16*, 1899. [CrossRef] [PubMed]
25. Hong, H.; Huang, C.; Ma, C.; Qi, J.; Shi, X.; Liu, C.; Wu, S.; Sun, Z.; Wang, E.; Liu, K. Twist-phase-matching in two-dimensional materials. *Phys. Rev. Lett.* **2023**, *131*, 233801. [CrossRef] [PubMed]
26. Tang, Y.; Sripathy, K.; Qin, H.; Lu, Z.; Guccione, G.; Janousek, J.; Zhu, Y.; Hasan, M.M.; Iwasa, Y.; Lam, P.K.; et al. Quasi-phase-matching enabled by van der Waals stacking. *Nat. Commun.* **2024**, *15*, 9979. [CrossRef]
27. Trovatello, C.; Ferrante, C.; Yang, B.; Bajo, J.; Braun, B.; Peng, Z.H.; Xu, X.; Jenke, P.K.; Ye, A.; Delor, M.; et al. Quasi-phase-matched up- and down-conversion in periodically poled layered semiconductors. *Nat. Photonics* **2025**, *19*, 291–299. [CrossRef]
28. Dai, D.; Tang, Y.; Bowers, J.E. Mode conversion in tapered submicron silicon ridge optical waveguides. *Opt. Express* **2012**, *20*, 13425–13439. [CrossRef]
29. Liu, J.; Duan, J.; Zhu, P.; Xia, G.; Hong, Q.; Zhang, K.; Zhu, Z.; Qin, S.; Xu, P. Modal phase-matching in thin-film lithium niobate waveguides for efficient generation of entangled photon pairs. *Opt. Express* **2024**, *32*, 40629–40639. [CrossRef]
30. Wang, C.; Xiong, X.; Andrade, N.; Venkataraman, V.; Ren, X.F.; Guo, G.C.; Lončar, M. Second harmonic generation in nano-structured thin-film lithium niobate waveguides. *Opt. Express* **2017**, *25*, 6963–6973. [CrossRef]
31. Imbrock, J.; Wesemann, L.; Kroesen, S.; Ayoub, M.; Denz, C. Waveguide-integrated three-dimensional quasi-phase-matching structures. *Optica* **2020**, *7*, 28–34. [CrossRef]
32. Nitiss, E.; Hu, J.; Stroganov, A.; Brès, C.S. Optically reconfigurable quasi-phase-matching in silicon nitride microresonators. *Nat. Photonics* **2022**, *16*, 134–141. [CrossRef]
33. Wang, J.F.; Yan, F.F.; Li, Q.; Liu, Z.H.; Liu, H.; Guo, G.P.; Guo, L.P.; Zhou, X.; Cui, J.M.; Wang, J.; et al. Coherent Control of Nitrogen-Vacancy Center Spins in Silicon Carbide at Room Temperature. *Phys. Rev. Lett.* **2020**, *124*, 223601. [CrossRef] [PubMed]
34. PsiQuantum Team. A manufacturable platform for photonic quantum computing. *Nature* **2025**, *641*, 876–883. [CrossRef]

35. Guo, Q.; Qi, X.Z.; Zhang, L.; Gao, M.; Hu, S.; Zhou, W.; Zang, W.; Zhao, X.; Wang, J.; Yan, B.; et al. Ultrathin quantum light source with van der Waals $NbOCl_2$ crystal. *Nature* **2023**, *613*, 53–59. [CrossRef]
36. Pelgrin, V.; Yoon, H.; Cassan, E.; Sun, Z. Hybrid integration of 2D materials for on-chip nonlinear photonics. *Light Adv. Manuf.* **2023**, *4*, 311–333. [CrossRef]
37. Azimi, A.; Barrier, J.; Barreda, A.; Bauer, T.; Bouzari, F.; Brokkelkamp, A.; de Mongeot, F.B.; Parsons, T.; Christianen, P.; Conesa-Boj, S.; et al. Photonics in Flatland: Challenges and Opportunities for Nanophotonics with 2D Semiconductors. *arXiv* **2025**, arXiv:2507.00336. [CrossRef]
38. Gyger, S.; Zichi, J.; Schweickert, L.; Elshaari, A.W.; Steinhauer, S.; Covre da Silva, S.F.; Rastelli, A.; Zwiller, V.; Jöns, K.D.; Errando-Herranz, C. Reconfigurable photonics with on-chip single-photon detectors. *Nat. Commun.* **2021**, *12*, 1408. [CrossRef]
39. Zhai, L.; Löbl, M.C.; Jahn, J.P.; Huo, Y.; Treutlein, P.; Schmidt, O.G.; Rastelli, A.; Warburton, R.J. Large-range frequency tuning of a narrow-linewidth quantum emitter. *Appl. Phys. Lett.* **2020**, *117*, 083106. [CrossRef]
40. Uppu, R.; Midolo, L.; Zhou, X.; Carolan, J.; Lodahl, P. Quantum-dot-based deterministic photon–emitter interfaces for scalable photonic quantum technology. *Nat. Nanotechnol.* **2021**, *16*, 1308–1317. [CrossRef]
41. Spinnler, C.; Nguyen, G.N.; Wang, Y.; Zhai, L.; Javadi, A.; Erbe, M.; Scholz, S.; Wieck, A.D.; Ludwig, A.; Lodahl, P.; et al. A single-photon emitter coupled to a phononic-crystal resonator in the resolved-sideband regime. *Nat. Commun.* **2024**, *15*, 9509. [CrossRef]

photonics

MDPI

Review

Silicon-Based On-Chip Light Sources: A Review

Yongqi Yang [1], Jiaqi Yang [1], Zhouyang Cheng [1], Shuyan Zhang [1], Zhen Yang [2], Shengchuang Bai [1] and Rongping Wang [1,*]

1 Laboratory of Infrared Material and Devices, Advanced Technology Research Institute, Ningbo University, Ningbo 315211, China
2 Suzhou Institute of Nano-Tech and Nano-Bionics, Chinese Academy of Sciences, Suzhou 215123, China
* Correspondence: wangrongping@nbu.edu.cn

Abstract

Silicon-based on-chip light sources are important since they can provide a compact solution for various applications in the field of high-speed optical communications, high-precision sensing, quantum information processing, and so on. We review the progress of silicon-based on-chip light sources in various materials. We provide some key parameters like pump thresholds, output powers, and pump schemes of on-chip lasers based on various materials. Finally, we point out the existing issues in the current investigations and possible solutions in the future.

Keywords: electronical pump; optical pump; waveguide; on-chip laser

1. Introduction

Emerging applications, such as big data, 5G networks, artificial intelligence, and cloud computing, have posed unprecedented challenges to the processing speed of signals, power consumption of systems, and transmission bandwidth of signals in current integrated optoelectronic devices. Compared with traditional telecom systems, all-optical signal processing, with the advantages of high transmission speed, broad bandwidth, electromagnetic interference immunity, low power consumption, and superior security, has gained significant attention in this high-speed optical communication era. Therefore, there is an urgent need to develop new-generation photonic integrated chips with high integration density, low power consumption, and efficient all-optical operation [1–4].

Silicon photonics is considered a leading platform for integrated optics due to its compatibility with existing CMOS manufacturing processes, which allows for cost-effective, large-scale production of photonic integrated circuits (PICs). Various materials have been deposited on silicon wafers for the development of various functional devices [1–4]. Current silicon-based optical components, including passive devices like optical waveguides, filters, and multiplexers, as well as active devices such as optical amplifiers [5], high-speed optical modulators [6], optical switches [7], and photodetectors [8], have attracted extensive attention in new-generation optical communication technologies. However, the slow development of on-chip silicon light sources has hindered the realization of large-scale integration for silicon photonic devices. This remains a major challenge in telecom systems. As an indirect bandgap semiconductor material [9–11], silicon exhibits low luminous efficiency. Moreover, silicon's 1.1 eV bandgap limits emitted light wavelengths to <1.1 μm. Clearly, developing on-chip light sources directly from silicon faces significant limitations.

To address these challenges, researchers have developed various on-chip light generation methods on silicon platforms, including stimulated Raman scattering [12], hybrid

https://doi.org/10.3390/photonics12070732

III-V/Si bonding [13], and rare-earth-doped gain media [14]. Raman-based approaches are constrained by silicon's narrow Raman gain bandwidth and strong two-photon absorption. III-V/Si heterogeneous integration represents a primary direction for silicon light sources, but suffers from limitations including short carrier lifetime, refractive index variations induced by electron–hole pair excitation, strong nonlinear optical responses, and spatiotemporal gain pattern effects caused by low emission efficiency in extended near/mid-infrared wavelengths. Moreover, the expensive and complex integration processes required for III-V semiconductors with silicon devices hinder their compatibility with standard CMOS processes.

In contrast, rare-earth (RE)-doped optical gain materials have recently shown significant promise for silicon-based light sources. These materials can be either doped into silicon or monolithically deposited on silicon, enabling low-cost wafer-scale manufacturing and universal device design. They exhibit advantages including high thermal stability, narrow intrinsic linewidth, and broadband optical gain/lasing through optical pumping [15,16]. However, silicon's low rare-earth doping concentration and high optical loss limit its luminous efficiency. An alternative effective approach involves integrating rare-earth-doped thin-film materials or gain devices with silicon chips. Erbium-doped systems, for instance, achieve population inversion through optical pumping and emit at 1.53 μm wavelength—corresponding to the minimum absorption window of silica fibers in optical communications. This wavelength remains unaffected by pump power or environmental temperature, serving as the standard wavelength in silicon photonics that meets communication-band signal processing requirements. Erbium-doped thin-film fabrication processes are compatible with silicon photonic manufacturing, offering attractive cost-effectiveness and design flexibility. Furthermore, combining population inversion in gain films with resonant enhancement in on-chip microcavities could ultimately enable on-chip laser emission. Therefore, developing novel erbium-integrated on-chip light sources compatible with silicon photonic systems represents an excellent solution to enhance silicon photonics' applicability in optical communications, holding significant importance for the advancement of silicon photonic technology.

In the present review, we concentrated on various on-chip lasers that can be integrated with silicon wafers. Especially, we focus on those electrically- and optically-pumped light sources with emission wavelengths around the telecom band. While the optically pumped light sources are dominant in most of the literature, electrically pumped sources may be more practical in the applications of integrated photonic devices. For other kinds of on-chip light sources, the readers can refer to other excellent review papers in optical frequency combs [17], supercontinuum sources [18], Raman lasers [19], and so on.

2. Hybrid III-V/Si Bonding Laser

It is challenging to realize high-efficiency light emission in silicon due to its indirect bandgap, leading to difficulty in achieving direct interband transitions and efficient radiative recombination of carriers. To overcome this physical constraint, researchers focus on direct bandgap semiconductor materials, like III-V compound semiconductors combining group III elements (Al, Ga, In) with group V elements (P, As, Sb). These materials exhibit multiple critical advantages in laser applications. Firstly, their intrinsic direct bandgap enables over 90% electron–hole radiative recombination efficiency, achieving orders-of-magnitude improvement in quantum efficiency compared to indirect bandgap materials like silicon. Secondly, precise control of ternary (e.g., InGaAsP) or quaternary (e.g., AlGaInAs) alloy compositions allow flexible bandgap engineering across 0.4–4.5 μm, meeting broad-spectrum applications from visible to mid-infrared. Moreover, exceptional high-frequency characteristics (carrier mobility > 3000 $cm^2/(V{\cdot}s)$, saturation velocity > 1 × 10^7 cm/s) support high-speed modulation beyond 10 Gb/s, fully compatible with modern optical communication standards. Furthermore, heterostructure designs such as quantum wells and quantum dots enable three-dimensional localization of carriers and optical modes, reducing threshold current density to the μA level while significantly enhancing thermal stability (characteristic temperature T_0 > 120 K). These properties establish III-V compounds as the cornerstone material system for high-performance lasers. Current research prioritizes developing silicon-compatible integration technologies, including low-temperature bonding and selective epitaxy, to achieve scalable photonic integration [20–22].

Various materials, including semiconductor quantum wells, nanowires, and quantum dots, have been deposited, and the corresponding devices have been fabricated for the development of on-chip lasers; below are the typical results.

2.1. Quantum Well and Nanowire

L. Cerutti et al. explored the viability of GaSb-based lasers for monolithic silicon integration toward 1.55 μm emission. By designing an active region based on strained $Ga_{0.8}In_{0.2}Sb$ quantum wells, effective carrier confinement was achieved. Edge-emitting lasers fabricated on GaSb substrates demonstrated continuous-wave operation at 1.56 μm up to 318 K. The same structure grown on silicon substrates exhibited room-temperature pulsed operation at 1.55 μm, achieving a threshold current density of 5 kA/cm^2 and an output power of 12 mW at 288 K under 200 mA driving current [23].

Dhruv Saxena et al. first demonstrated room-temperature lasing in GaAs/AlGaAs/GaAs core-shell-cap nanowires through precise Fabry-Pérot cavity design. Under 522 nm optical pumping, the device exhibited threshold pump fluence of 96 μJ/cm^2 at 6 K and 207 μJ/cm^2 at 300 K, with a threshold gain of 1820 cm^{-1}. This breakthrough marks a critical step toward integrating GaAs nanowire lasers into near-infrared nanophotonic devices [24].

Jun Wang et al. reported an InGaAs/AlGaAs quantum well laser structure prepared by metal–organic chemical vapor deposition (MOCVD). Material characterization revealed that this method effectively reduces threading dislocation density and interfacial roughness, allowing the optical performance of the silicon-based active region to approach levels achieved in structures grown on native GaAs substrates. Fabricated broad-area lasers (1 mm cavity length, 15 μm stripe width) demonstrated an exceptionally low threshold current density (313 A/cm^2) under room-temperature pulsed operation. The laser achieved continuous-wave operation at 240 K [25].

Alexander Spott et al. reported the first heterogeneously integrated 4.8 μm quantum cascade laser (QCL) on a silicon platform, utilizing an InP-based InGaAs/InAlAs quantum cascade active region integrated with a silicon-on-nitride-on-insulator (SONOI) waveguide structure (as shown in Figure 1a). The device showed room-temperature pulsed operation with a threshold current of 387–388 mA, maximum output power of 31 mW, and slope efficiency of 170 mW/A at 293 K (as shown in Figure 1b). Future optimizations in thermal management and waveguide design may enable continuous-wave operation [26].

Meixin Feng et al. proposed a preliminary on-chip integration scheme featuring GaN-based lasers, modulators, and photodetectors monolithically grown on silicon substrates. The modulator, integrated and sharing the same InGaN quantum well active region with the laser, enabled light intensity control by modulating applied biases to induce energy band bending in the active region, thereby adjusting absorption characteristics. This approach achieved threshold current and optical power modulation, with a spectral linewidth of 1.0 nm and a tunable threshold current (minimum at +2 V and increased to 770 mA at −4 V) accompanied by a slope efficiency of 98.1 mW/A. The photodetector effectively monitored the modulated laser output, demonstrating the feasibility of voltage-tunable intra-system light detection [27].

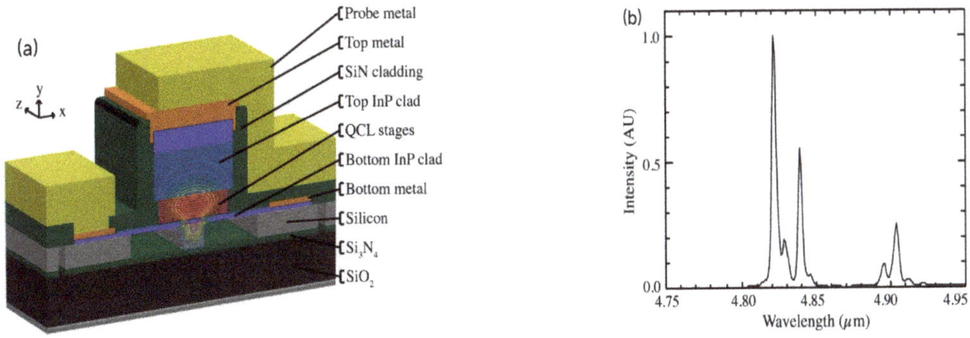

Figure 1. (**a**) Cross-sectional schematic illustrating the hybrid silicon–QCL active region structure, overlaid with a contour plot of the simulated fundamental TM optical mode's electric field component. (**b**) Emission spectra of the device measured with a monochromator at 293 K. Reprinted with permission from [26].

The team subsequently reported a room-temperature electrically pumped InGaN-based microdisk laser grown on silicon substrate. The device, fabricated via AlGaN cladding layers (vertical optical confinement factor 1.84%) and microring resonator configuration (R = 20 µm/r = 10 µm), achieved lasing at 412.4 nm under 250 mA threshold current with spectral linewidth narrowed to 0.4 nm, as shown in Figure 2a,b. Electroluminescence spectra exhibited distinct slope discontinuity in light–current curves in Figure 2c, providing conclusive evidence of lasing. This work marks the first observation of room-temperature electrical pumping in InGaN-based microdisk lasers directly grown on silicon substrates [28].

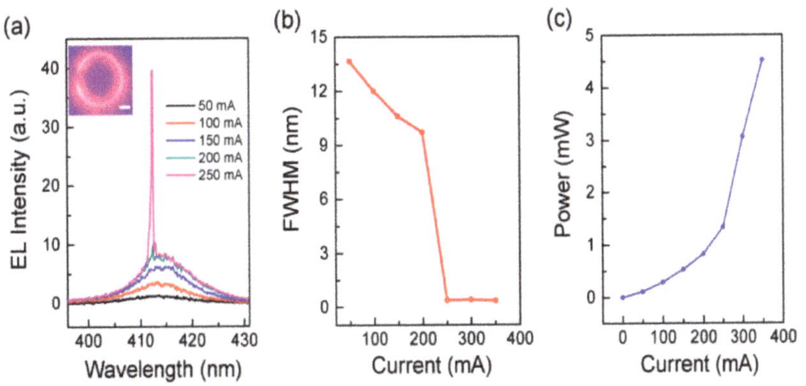

Figure 2. (**a**) EL spectra of an InGaN-based microring laser grown on Si (R = 20 µm and r = 10 µm) measured under various pulsed currents. The inset shows a top-view emission pattern of the device at a pulsed injection current of 200 mA (below the lasing threshold), and the scale bar is 10 µm. (**b**) FWHM of the EL spectra as a function of the pulsed injection current. (**c**) EL light output power as a function of the injection current. Reprinted with permission from [28].

Hoang Nguyen-Van et al. reported QCLs directly grown on silicon substrates with an emission at 11 µm. The devices exhibited a room-temperature (300 K) threshold current density as low as 1.3 kA/cm^2 (3 mm cavity length) with differential gain of 16.9 cm/kA and maximum operating temperature exceeding 380 K, matching the performance of InAs-substrate counterparts (threshold current density 1.03 kA/cm^2) [29].

Z. Loghmar et al. demonstrated monolithically integrated indium arsenide/aluminum antimonide (InAs/AlSb) QCLs on (001) silicon substrates. By utilizing a 1.5-μm-thick GaSb buffer layer and high-temperature substrate pretreatment (1000 °C annealing), antiphase-domain-free epitaxial growth was realized on silicon substrates. The lasers emitted at 8 μm wavelength under room-temperature (300 K) operation, with threshold current densities of 0.92–0.95 kA/cm^2 for 3.6 mm-long device in pulsed mode. Comparative studies showed that QCLs with identical designs grown on native InAs substrates exhibited lower threshold current densities (0.75 kA/cm^2) while maintaining equivalent maximum operating temperatures as shown in Figure 3. The low threshold current density characteristics of these silicon-based QCLs make them particularly suitable for photonic integrated sensor applications [30]. The relatively high threshold current density in the silicon-based InAs/AlSb QCL is primarily attributed to crystal defects introduced during heteroepitaxial growth. High dislocation density resulting from the lattice mismatch between the silicon substrate and III-V materials significantly increases non-radiative recombination and reduces carrier efficiency. Additionally, the rougher heterointerface of the silicon-based material (RMS reaching 9.9 nm) increases optical scattering losses, while material inhomogeneity slightly broadens the gain spectrum, collectively pushing up the threshold current.

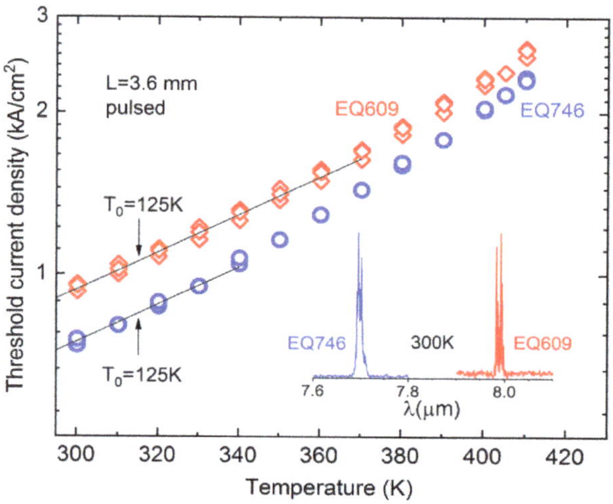

Figure 3. Threshold current density versus temperature (semi-logarithmic scale) for QCLs grown on silicon (EQ609, ridge widths: 13 μm, 14 μm, and 14.5 μm) and on InAs substrates (EQ746, ridge widths: 13 μm and 17 μm). The inset displays emission spectra measured at 0.6 A and 300 K. Reprinted with permission from [30].

2.2. Quantum Dots

Recent research on quantum dot lasers based on the InAs/GaAs material system has garnered significant attention, with scientific efforts focusing on structural design, fabrication processes, and optoelectronic performance optimization [31–35]. Notably, Chen et al. achieved the first electrically driven continuous-wave InAs/GaAs quantum dot laser. Using metal–organic chemical vapor deposition (MOCVD), a 400-nm-thick anti-phase boundary-free GaAs film with a root-mean-square (RMS) surface roughness as low as 0.86 nm was epitaxially grown on silicon substrates, forming a high-quality virtual substrate. Subsequent molecular beam epitaxy (MBE) enabled the fabrication of quantum dot laser structures on this GaAs/Si (001) platform, delivering breakthrough performance: under room-temperature continuous-wave operation, the laser demonstrated a threshold current density as low as 425 A/cm^2 with single-facet output power reaching 43 mW at a stable emission wavelength of 1.3 μm. In pulsed mode, it achieved a threshold current density to 250 A/cm^2 and single-facet output power surpassing 130 mW at room temperature, and the device worked even at a temperature of 375 K [33].

Liu et al. reported the electrically pumped continuous-wave (CW) III-V semiconductor laser monolithically grown on (001) silicon substrate. This achievement was realized using InAs/GaAs quantum dots as the active region and incorporating a GaP buffer layer between the silicon substrate and device layer. The laser had a threshold current density of ~860 A/cm^2 and single-facet output power of 110 mW. This work validates the compatibility of high-performance III-V light sources with standard silicon substrates, providing a critical technical pathway for silicon photonic integration [34].

Chen Shang et al. demonstrated the electrically pumped quantum dot laser grown via molecular beam epitaxy on 300-mm patterned (001) silicon wafers, employing InAs/GaAs quantum dots as the gain medium combined with a buffer layer architecture based on AlGaAs/GaAs material systems. At 20 °C, the maximum double-side output power reaches 126.6 mW, and a threshold current is 47.5 mA in Figure 4a. The device operates in continuous-wave (CW) mode up to 60 °C, as shown in Figure 4b, exhibiting a distinct lasing peak at 1300 nm above the threshold in Figure 4c. Figure 4d shows minimal variation in the bias voltage required to maintain 10 mW output power over 350 h, indicating device reliability approaching that of similar lasers grown on bare silicon substrates. This heterogeneous epitaxial integration platform on silicon holds the potential for robust on-chip laser operation and efficient low-loss coupling to silicon photonic circuits, offering prospects for scalable and low-cost mass production [35].

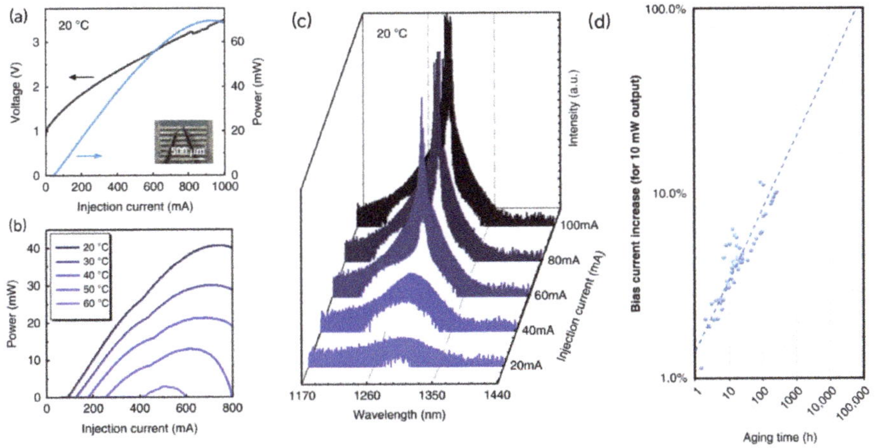

Figure 4. Laser measurement results. (**a**) Room-temperature light–current–voltage (LIV) of the device with the highest output power. The inset depicts probe needles contacting a cleaved laser bar facet. (**b**) Temperature-dependent LIV shows lasing up to 60 °C. (**c**) Room-temperature lasing spectrum evolution versus injection current. (**d**) Aging results for the in-pocket laser at 35 °C. Reprinted with permission from [35].

2.3. InP

InP-based semiconductor lasers demonstrate superior device performance in the near-infrared (1.3–1.6 μm) band, characterized by low threshold current density and high energy conversion efficiency, owing to their direct bandgap characteristics, high carrier mobility, and exceptional optical field confinement.

Recent studies [36–47] have achieved breakthrough improvements in device performance through heterostructure design and interface engineering optimization. For example, Wang et al. developed a III-V/Si hybrid integrated single-microcavity narrow-linewidth laser combining an InP multiple quantum well gain section with silicon-based microring resonators (Si/SiN waveguides). The device achieved single-mode lasing with a side-mode suppression ratio (SMSR) exceeding 45 dB; the peak value reaches 48 dB, as shown in Figure 5, and a maximum output power of 16.4 mW. Compared to current hybrid/heterogeneous integrated laser architectures requiring multiple complex controls, this design simplifies laser characterization complexity for mass production through only two control parameters. This streamlined structure and control framework simultaneously realized an ultranarrow linewidth of 2.79 kHz and a low relative intensity noise (RIN) of −135 dB/Hz. The system demonstrated a 10 dB signal-to-noise ratio (SNR) in 12.5 Gb/s optical data transmission experiments, showing potential for phase noise-sensitive applications such as coherent optical communications and LiDAR systems [38].

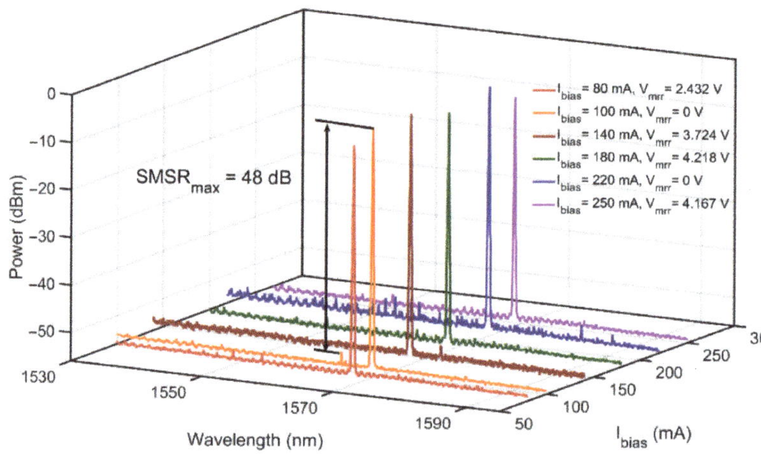

Figure 5. Laser evolution spectrum at different I_{bias} and V_{mrr} settings. Reprinted with permission from [38].

Chao Xiang et al. demonstrated optical amplification via current injection in a hybrid InP/Si gain waveguide, combined with an extended distributed Bragg reflector (E-DBR) design: one side features a 20 mm long SiN grating (providing narrowband feedback with κ values of 0.25–0.875 cm^{-1}) in Figure 6a,b, while the opposite side integrates a silicon-based Mach–Zehnder interferometer (MZI) broadband tunable reflector, forming a hybrid integrated external-cavity laser. For a laser operating at telecommunication wavelengths (1548 nm in Figure 6c), Figure 6d reveals mode hopping induced by intracavity phase variations in the laser; this laser delivers >10 mW output power through SiN waveguides, as shown in Figure 6a, achieving threshold current densities as low as 580 A/cm^2, baseband linewidths <1 kHz in Figure 6e, and relative intensity noise (RIN) < −150 dBc/Hz (up to 20 GHz) in Figure 6f. Through self-injection locking coupled with a SiN microring resonator (Q-factor of 42 million), the frequency noise is reduced by 20–30 dB, realizing a Lorentzian linewidth of 3 Hz [46].

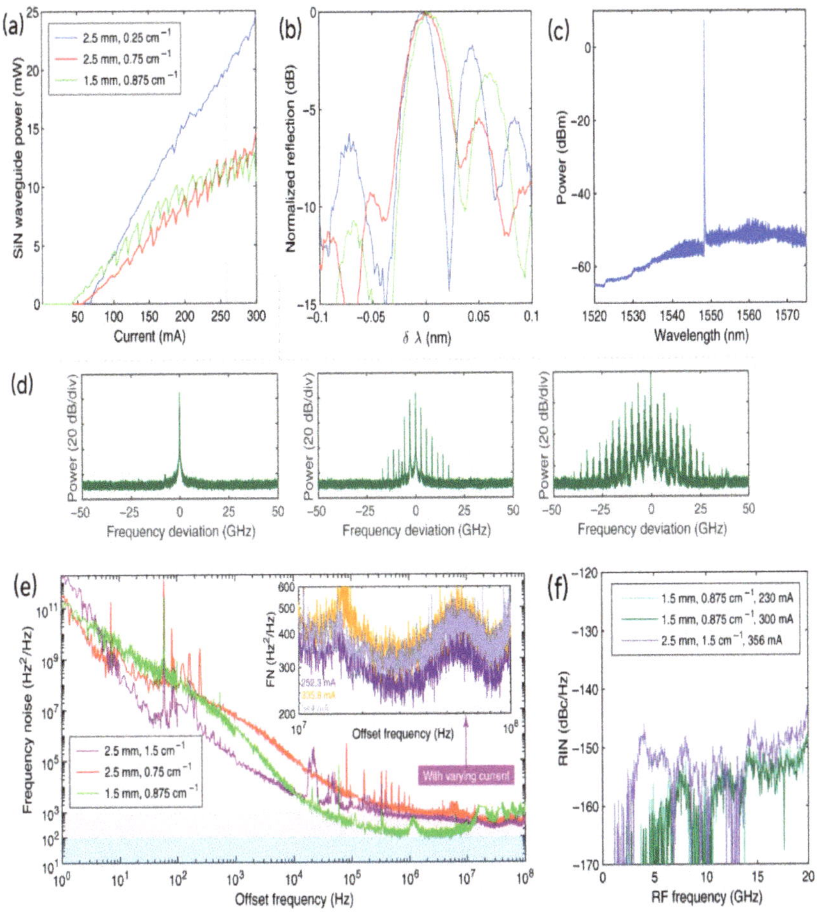

Figure 6. Laser characterization. (**a**) LI measurements of three lasers with different gain section length (1.5 mm or 2.5 mm) and grating κ values (0.25 cm^{-1}, 0.75 cm^{-1}, and 0.875 cm^{-1}). (**b**) Corresponding grating reflection spectra of the lasers in (**a**). (**c**) Optical spectrum of the high-power laser (shown in blue in (**a**,**b**)) at a gain current of 300 mA. Optical spectrum analyzer operated at 0.02 nm resolution bandwidth. (**d**) Single-mode/multimode transition during DBR laser wavelength red-tuning through mode-hop cycles. The corresponding laser gain currents from left to right are 254.4, 258, and 261 mA, respectively. (**e**) Laser frequency noise spectra exhibiting low-frequency noise characteristics with fundamental linewidths at kHz and sub-kHz levels. The laser currents are 252.3 mA (purple), 382.2 mA (red) and 298 mA (green). The inset shows current-dependent high-offset frequency noise spectra for the laser with 2.5 mm gain section length and 1.5 cm^{-1} grating κ value (purple), indicating that sub-kHz level Lorentzian linewidth is achieved for all three currents. (**f**) Relative intensity noise (RIN) spectra for a low-threshold laser (green traces in (**a**,**b**)) and a reference device (purple trace in (**e**)). Reprinted with permission from [46].

Table 1 lists the typical results of the III-V-based semiconductor laser integrated with silicon, including emission wavelength, threshold pump power, and maximal output power. In most cases, the output powers are limited at around few or tens of mW; for practical applications, it is highly desired to have a high-power laser output with relatively lower pump power. Especially for the on-chip applications, it might be more important to have a laser with low pump power and high conversion ratio of pump to output power.

Table 1. Summary of the typical results in various III-V-based semiconductor lasers.

Material	Threshold (Threshold Intensity)	Working Wavelength	Maximum Output Power	Pumping Scheme	Ref.
GaSb	5000 A/cm^2	1550 nm	12 mW	Electrical	[23]
GaAs	207 μJ /cm^2 per pulse	830–870 nm	/	Optical	[24]
GaAs	313 A/cm^2	976.8 nm	>300 mW	Electrical	[25]
InP	0.9 mA	1526–1532 nm	0.93 mW	Electrical	[36]
InP	22 ± 2.1 mW	930.5 nm	6.4 mW	Optical	[37]
InP	52 mA	1563–1566 nm	16.4 mW	Electrical	[38]
InAs/GaAs	650 A/cm^2	1290 nm	/	Electrical	[31]
InAs/GaAs	62.5 A/ cm^2	1310 nm	>105 mW	Electrical	[32]
InAs/GaAs	425 A/cm^2	1285–1292 nm	43 mW	Electrical	[33]
InAs/GaAs	860 A/cm^2	1280 nm	110 mW	Electrical	[34]
InAs/GaAs	47.5 mA	1300 nm	126.6 mW	Electrical	[35]
InGaAs/InAlAs	387–388 mA	4800 nm	31 mW	Electrical	[26]
GaN	700 mA	405 nm	/	Electrical	[27]
InGaN	250–1600 mA	412.4 nm	/	Electrical	[28]
InAs/AlSb	920–950 A/cm^2	8000 nm	/	Electrical	[30]
InAs/AlSb	1300 A/cm^2	1050–1110 nm	>100 mW/facet	Electrical	[29]
InP	36 mA	1310 nm	22 mW	Electrical	[39]
InP	17 mA	1566 nm	6 mW	Electrical	[40]
InP	17 mA	1566 nm	6 mW	Electrical	[41]
InP	100 μA	1560 nm	95 ± 2 μW	Electrical	[42]
InP	51 mA	1544 nm	/	Electrical	[43]
InP	100 mA	1480–1590 nm	3 mW	Electrical	[44]
InP	50 μJ/cm^2 per pulse	870–910 nm	/	Optical	[45]
InP	42 mA	1548 nm	>10 mW	Electrical	[46]
InP	30 mA	1541–1589 nm	/	Electrical	[47]

3. Al_2O_3

Al_2O_3 is probably most investigated host materials for the development of rare-earth ions doped on-chip lasers. Several important material properties, like low propagation loss, good rare-earth ion solubility, and high gain characteristics, make Al_2O_3 an ideal platform for the development of RE-doped on-chip lasers [48]. The Al_2O_3-based on-chip lasers can be categorized into different types: microring resonator-based lasers and distributed Bragg reflector-based lasers, including distributed feedback lasers (DFB); and distributed Bragg reflector lasers (DBR).

3.1. Microring Resonator Lasers

Microring resonator lasers offer advantages such as compact size (typically on the order of tens of micrometers), high integration density, high quality factor (Q values up to 10^5–10^6), low threshold, and CMOS compatibility. Starting from the initial observation of net optical gain in Al_2O_3:Er^{3+} thin films by Polman's team in 1996 [49], and after over a decade of exploration, Bradley's team achieved on-chip lasing in 2009 using a microring resonator structure, demonstrating the first low-threshold laser emission at communication wavelengths on a silicon platform [50]. This breakthrough provided a CMOS-compatible light source solution for photonic integrated circuits. The device balanced high pump coupling efficiency and low signal output coupling through optimized microring resonator design. A 500-nm-thick Al_2O_3:Er^{3+} layer was deposited on a silicon substrate via reactive co-sputtering, followed by photolithography and reactive ion etching to form 1.5-μm-wide single-mode waveguides, capped with a 5-μm-thick SiO_2 upper cladding layer. By adjusting the directional coupler length (350–600 μm) and microring cavity length (2.0–5.5 cm), broad wavelength selectivity could be achieved, as shown in Figure 7. Within the 1530–1557 nm range, the lasing wavelength was tuned by modifying the output coupling ratio. The device exhibited a maximum output power of 9.5 μW, a slope efficiency of 0.11%, and a threshold pump power as low as 6.4 mW, outperforming previously reported Er^{3+}-doped microring lasers. Bradley's team further explored optimization strategies such as ytterbium (Yb^{3+}) co-doping, unidirectional output structures, and distributed feedback to enhance device efficiency, reduce size, and achieve single-frequency operation.

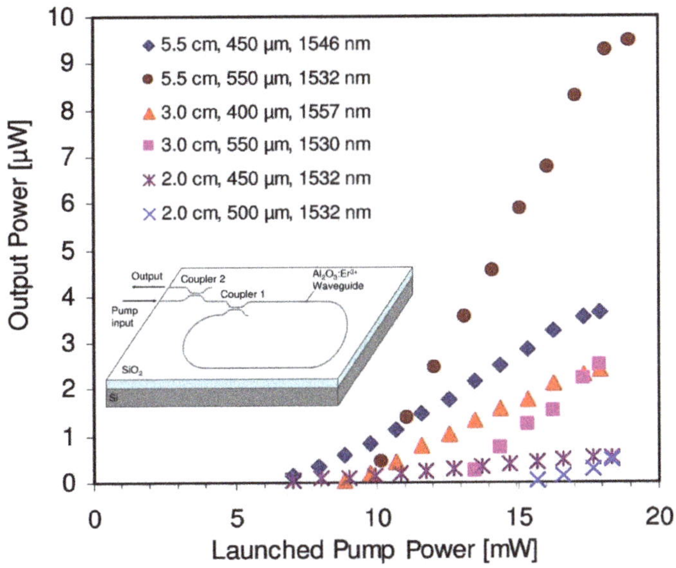

Figure 7. On-chip laser output vs. pump power for varying resonator and output coupler lengths. Resonator length, coupler length, and main lasing wavelength are specified. The inset in Figure 7 is a schematic of an Al_2O_3:Er^{3+} ring laser. Reprinted with permission from © Optical Society Group [50].

In 2014, Bradley et al. demonstrated a monolithically integrated erbium (Er^{3+})/ytterbium (Yb^{3+})-doped microring laser based on a silicon-compatible process. The laser employs a dual-SiN-layer structure, with a microring outer diameter of 160 µm and an Al_2O_3:Er^{3+} gain layer thickness of 2 µm. For the undoped microring, the measured maximum intrinsic quality factors (Q_i) were 3.8×10^5 at 980 nm and 5.7×10^5 at 1550 nm. The Er^{3+}-doped device achieved single-mode lasing at 1559.82 nm with a threshold of 0.5 mW, a double-ended slope efficiency of 0.3%, and a maximum output power of 2.4 µW. The Yb^{3+}-doped device demonstrated single-mode operation at 1042.74 nm, exhibiting a threshold of 0.7 mW, a slope efficiency of 8.4%, and an output power exceeding 100 µW. The devices are coupled via 4 µm deep trenches to SiN waveguides with widths of 0.4/0.9 µm and optimized gap dimensions of 0.3–0.5 µm. This design enables high-density silicon-based on-chip integration through a SiN-Al_2O_3 composite cavity structure, reducing the device footprint by 500 times and lowering the threshold by over an order of magnitude compared to traditional waveguide lasers. It provides a scalable micro/nano light source solution for optical communications (1.5 µm) and biophotonics (1.0 µm) applications [51].

In 2023, Wang et al. reported a photonic chip-integrated titanium–sapphire (Ti:Sa) laser by combining Ti:Sa gain media with a silicon nitride-sapphire integrated photonic platform. The research team designed a microring resonator with a 200 µm radius, achieving laser emission across the 730–830 nm wavelength range. By confining both pump and lasing modes within a single microring, the device reduced the threshold power to 6.5 mW—two orders of magnitude lower than conventional Ti:Sa lasers. Single-mode operation was realized through a distributed Bragg reflector (DBR) and feedback loop structure, yielding a side-mode suppression ratio (SMSR) of 16 dB, along with a 0.5 mW fiber-coupled output power and a narrow linewidth of 120 kHz [52].

In 2024, Yang et al. reported a photonic integrated platform based on monocrystalline titanium–sapphire-on-insulator (Ti:SaOI). Utilizing a 50 µm-diameter microring resonator, they achieved a Q factor of 1.3×10^6 and demonstrated continuous-wave laser emission with an ultralow threshold of 290 μW under 532 nm pumping. High-confinement waveguides enabled amplified spontaneous emission (ASE) with a peak gain of 64 dB/cm and a bandwidth of 100 THz. An 8 mm spiral waveguide achieved 20 dB total gain under 175 mW pumping, successfully amplifying 2.2 ps pulses without distortion to 1.0 kW peak power and 2.3 nJ energy. Integration of a platinum micro-heater enabled 50 nm wavelength tuning in an 80 μm microring, yielding 140 kHz linewidth and 1.8 mW output power when pumped by commercial laser diodes. The platform manipulated silicon carbide quantum emitters via laser arrays. This system reduces the size and cost of traditional Ti:sapphire setups by three orders of magnitude, achieves the first solid-state amplification in sub-micrometer bands, and provides miniaturized solutions for quantum optics and high-power frequency comb applications [53].

3.2. DBR and DFB

Generally, microring lasers are prone to multimode oscillation, requiring coupling gap optimization or auxiliary structures to maintain single-mode lasing. Additionally, limited by their small mode field volume, microring lasers typically exhibit low output power (e.g., microwatt-level), making them unsuitable for direct high-power applications. In contrast, DBR lasers demonstrate superior single-mode stability and higher output power capabilities.

In 2013, Purnawirman et al. utilized a wafer-scale immersion lithography process to fabricate grating structures combined with silicon nitride waveguides, forming a distributed Bragg reflector (DBR). A layer of erbium-doped aluminum oxide (Al_2O_3:Er^{3+}) was then deposited atop as the gain medium, creating an inverted ridge waveguide. The DBR cavity achieved an optical power output exceeding 5 mW, with laser emission observed across a broad wavelength range, including peaks at 1536 nm, 1561 nm, and 1596 nm [54].

Compared to DBR lasers, DFB lasers exhibit narrower linewidths and higher side mode suppression ratios (SMSR). These characteristics are critical for devices requiring long coherence lengths and extremely high phase sensitivity, such as those used in coherent optical communication systems. Michael Belt et al. reported an arrayed erbium-doped aluminum oxide (Al_2O_3:Er^{3+}) waveguide distributed feedback (DFB) laser based on an ultra-low-loss silicon nitride (Si_3N_4) platform. This laser employed 248 nm stepper lithography to define sidewall grating structures within the Si_3N_4 layer, combined with a reactive co-sputtering process to fabricate the erbium-doped gain layer, achieving a monolithically integrated design. A five-channel laser array with wavelengths distributed across 1531–1543 nm (spanning 12 nm) demonstrated a single-mode output power of 8 μW, a side mode suppression ratio (SMSR) exceeding 35 dB, and a linewidth of 501 kHz [55].

Subsequently, Michael Belt et al. further refined the integration of erbium-doped waveguide distributed Bragg reflector (DBR) and distributed feedback (DFB) laser arrays on an ultra-low-loss silicon nitride (Si_3N_4) waveguide platform. By optimizing cavity design and leveraging sidewall grating technology enabled by the ultra-low-loss platform, they achieved low threshold power and high slope efficiency. The DBR lasers exhibited a threshold as low as 11 mW under 974 nm pumping, with a pump-to-signal conversion efficiency of 5.2%. For the DFB lasers, extending the cavity length to 21.5 mm (nearly three times longer than previous designs) enhanced pump efficiency by over two orders of magnitude, reaching up to 0.77%. Figure 8 demonstrates the spectral performance of the DBR and DFB laser arrays: Figure 8a displays the output spectra of five DBR lasers with distinct grating periods (ranging from 478 nm to 486 nm), emitting at wavelengths of 1535 nm, 1541 nm, 1547 nm, 1554 nm, and 1560 nm, respectively; Figure 8b shows the output spectra of four DFB lasers with grating periods (from 478 nm to 490 nm), operating at wavelengths of 1534 nm, 1546 nm, 1558 nm, and 1570 nm; all devices exhibit side mode suppression ratios exceeding 50 dB, and the output power variations arise from the wavelength-dependent gain threshold and maximum small-signal gain spectrum of the erbium-doped gain layer. The devices demonstrated stable performance under uncooled conditions, and future improvements could further boost efficiency by adopting 1480 nm pumping to mitigate energy-level bottlenecks, introducing Yb sensitizers, or integrating pump filters. This technology offers a scalable solution for multi-wavelength array integration with high thermal stability and narrow linewidth [56].

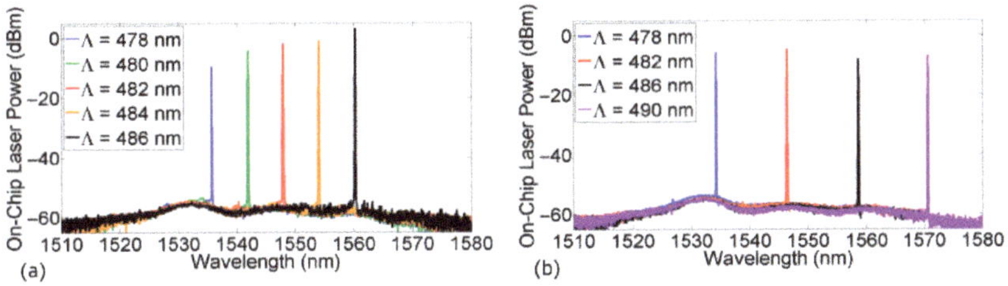

Figure 8. (**a**) Superimposed DBR output laser spectra. (**b**) Superimposed DFB output laser spectra. Reprinted with permission from [56].

Purnawirman et al. subsequently reported an erbium-doped aluminum oxide ($Al_2O_3:Er^{3+}$)-based distributed Bragg reflector (DBR) and distributed feedback (DFB) laser utilizing a multi-segment silicon nitride waveguide architecture. Employing a five-segment silicon nitride waveguide design, they embedded the waveguides in silicon dioxide (SiO_2) and integrated them with a top erbium-doped aluminum oxide layer to achieve lasing in two configurations: a DBR cavity with a 13.8 mm gain length (featuring 5 mm periodic sidewall gratings) and a DFB cavity with a 20 mm full-gain region (incorporating 550 nm spaced gratings). The DBR laser delivered a maximum output power of −3.6 dBm (0.44 mW) at 1565 nm, with a slope efficiency of 1.4% and a threshold pump power of 64 mW. The DFB laser, benefiting from grating optimization, achieved a lower threshold (14 mW), an output power of −7.3 dBm (0.18 mW), and an improved slope efficiency of 2.7%. By thickening the silicon nitride layer to 200 nm, this design overcame the integration limitations of conventional thin-layer waveguides while maintaining efficient mode confinement in the gain medium, paving the way for monolithic integration in communication, sensing, and multi-wavelength photonic systems [57].

In 2017, Purnawirman et al. reported a wavelength-insensitive waveguide design based on a silicon nitride platform, achieving ultra-narrow-linewidth erbium-doped aluminum oxide ($Al_2O_3:Er^{3+}$) distributed feedback (DFB) lasers. The waveguide consists of five segmented silicon nitride (SiNx) sections (200 nm thick, 450 nm wide, with 400 nm gaps) embedded in silicon dioxide (SiO_2), topped with a 1100 nm thick $Al_2O_3:Er^{3+}$ gain layer. The schematic diagram of the device structure is shown in Figure 9a. By comparing discrete quarter-phase-shifted (QPS-DFB) and distributed-phase-shifted (DPS-DFB) cavity configurations, the QPS-DFB lasers achieved on-chip output powers of 0.41 mW, 0.76 mW, and 0.47 mW at C/L-band wavelengths (1536 nm, 1566 nm, and 1596 nm), with slope efficiencies of 0.3–0.6% and threshold powers ranging from 55 mW to 105 mW. Figures 9b and 9c respectively illustrate the optical spectra at different grating periods and the on-chip output power versus pump characteristics of the distributed-phase-shifted distributed feedback (DPS-DFB) lasers. In contrast, the DPS-DFB laser demonstrated significantly enhanced performance at 1565 nm, delivering an output power of 5.43 mW (slope efficiency: 2.9%, threshold: 14 mW) and a side mode suppression ratio (SMSR) exceeding

59.4 dB. This work provides a critical reference for integrating narrow-linewidth lasers on silicon photonics platforms [58].

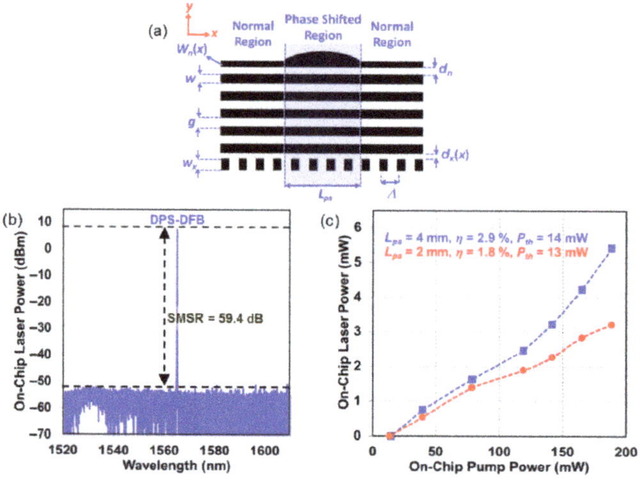

Figure 9. (**a**) Schematic of an Al_2O_3:Er^{3+} QPS-DFB laser cavity (not to scale). The design uses five continuous SiNx segments with grating perturbation introduced by two laterally positioned periodic structures. (**b**) Emission spectra of these lasers for different grating periods. (**c**) On-chip output laser power vs. pump power for these lasers. Reprinted with permission from [58].

Table 2 summarizes the current development of on-chip laser technologies utilizing RE-doped alumina materials, with key performance parameters covering laser wavelength, maximum slope efficiency, maximum output power, and linewidth. Microring resonator lasers feature miniaturized size, high integration density, and low threshold, yet their output power is typically below 10 μW due to the limitation of small mode volume; in contrast, DBR lasers combine single-mode stability with high-power output, while DFB lasers demonstrate unique advantages in phase-sensitive scenarios like coherent optical communications by offering ultra-narrow linewidth and high side-mode suppression ratio.

Table 2. Overview of on-chip lasers on rare-earth ion-doped Al_2O_3.

Cavity Type	Rare-Earth Dopant	Laser Wavelength	Maximal Slope Efficiency	Maximal Laser Output Power	Linewidth	Reference
Ring	Er/Yb	1560 nm/1042 nm	0.3%/8.4%	2.4 μW/100 μW	/	[51]
Ring	Tm	1760–1920 nm	24%	230 μW	/	[59]
Ring	Ti	790–930 nm	2.9%	1.8 mW	140 kHz	[53]
DFB	Er	1543 nm	0.006%	9 μW	501 kHz	[55]
DFB	Er	1560 nm	0.77%	270 μW	/	[56]
DFB	Er	1563 nm	2.2%	1.0 mW	/	[60]
DFB	Er	1563–1580 nm	/	20 μW	/	[61]
DFB	Er	1553 nm	1.3%	2.6 mW	/	[62]
DFB	Er	1592 nm	0.7%	1.2 mW	/	[63]
DFB	Er	1599 nm	0.01%	/	/	[64]
dps-DFB	Er	1565 nm	2.9%	5.43 mW	5.3 kHz	[58]
qps-DFB	Er	1536, 1566, 1596 nm	0.6%	0.76 mW	30.4 kHz	[58]
DFB	Tm	1861 nm	14%	267 mW	/	[65]
DFB	Ho	2022–2101 nm	2%	15 mW	/	[66]
DBR	Er	1536, 1561, 1596 nm	2.6%	5.1 mW	/	[54]
DBR	Er	1546 nm	5.2%	2.1 mW	/	[56]
DBR	Er	1564 nm	0.02%	/	/	[14]
DBR	Tm	1881 nm	23%	387 mW	/	[65]

4. $LiNbO_3$

As a representative electro-optic material, lithium niobate exhibits a strong electro-optic effect, enabling rapid dynamic modulation of laser wavelength, phase, and output intensity through external electric fields—a critical capability for designing tunable narrow-linewidth lasers. Additionally, its remarkable nonlinear optical properties support efficient frequency conversion processes (e.g., frequency doubling and parametric oscillation), facilitating multi-wavelength laser output on a single chip and expanding spectral coverage.

In 1992, P. Becker et al. pioneered the first Er-diffused Ti:$LiNbO_3$ single-mode waveguide laser, achieving continuous-wave and pulsed operations at 1576 nm (threshold: 13 mW under 1479 nm π-polarized pumping) and 1563 nm (activated at 25 mW). Subsequent advancements by J. Amin et al. demonstrated Nd-diffused and rare-earth-doped $LiNbO_3$ lasers, where optimizing cavity mirrors with 95% reflectivity reduced thresholds from 39 mW to 4.2 mW and achieved 8.5% slope efficiency at 1531.4 nm. J.K. Jone et al. reported Yb-doped waveguide lasers, addressing photorefractive instability via wet-oxygen annealing (500 °C, 14 h), enhancing a 14 mm device to 16% slope efficiency. Ingo Baumann et al. optimized Ti:Er:$LiNbO_3$ lasers theoretically, attaining 37% slope efficiency (1561 nm, 63 mW output at 210 mW pump), highlighting communication compatibility. S. BALSAMO et al. developed a Q-switched Ti:Er:$LiNbO_3$ laser, delivering 2.4 W peak power and 0.18 μJ pulses at 2 kHz under 100 mW pumping. These milestones underscore $LiNbO_3$-based lasers' potential for high-efficiency, stable, and pulsed applications in photonic technologies [67–72].

Das, BK and collaborators first demonstrated a single-frequency Ti:Er:$LiNbO_3$ distributed Bragg reflector (DBR) waveguide laser containing two thermally fixed photorefractive Bragg gratings. The laser initiated lasing at 70 mW pump power, achieving 1.1 mW output power at 120 mW pump power with a slope efficiency of approximately 2%. Operating in TE-polarized single longitudinal mode, it exhibited a linewidth of ~0.75 GHz. Temperature adjustment of the output coupling grating enabled a 80 pm tuning range with a tuning rate of 8 pm/°C. Experiments confirmed that resonator optimization could enhance output power and slope efficiency, suggesting thermal tuning and electro-optic tuning as promising methods for developing tunable single-frequency light sources [73].

R. E. Di Paolo and colleagues reported continuous-wave laser operation in Zn-diffused $LiNbO_3$:Nd^{3+} channel waveguides at room temperature. Measured in a 10 μm wide, 0.9 cm long waveguide, the laser exhibited a threshold pump power of 1.25 mW and a slope efficiency of 20%. Sustained stable emission without photorefractive damage was achieved for over 20 min under maximum pump power, reaching a maximum output power of 0.14 mW. The laser operated in TM polarization, providing a novel approach for developing $LiNbO_3$-based waveguide lasers [74].

Masatoshi Fujimura et al. first demonstrated a Yb-diffused $LiNbO_3$ annealed/proton-exchanged (APE) waveguide laser with a wavelength bandwidth of 0.36 nm. The device achieved stable continuous-wave lasing at 1061 nm under 918 nm pumping. For a 5 μm channel-width device, the threshold pump power was 40 mW, delivering a maximum output power of 1.2 W with a slope efficiency of 3×10^{-5} [75].

Fujimura et al. proposed and demonstrated a Yb:$LiNbO_3$ APE waveguide laser pumped by a 980 nm InGaAs laser diode, achieving lasing at 1061 nm. The device exhibited a spectral bandwidth of 0.3 nm, a threshold pump power of 30 mW, and an output power of 4.3 mW at 90 mW pump power—three orders of magnitude higher than previous reports. This marked a significant advancement in output power, offering critical insights for related fields [76]. In the waveguide with an optimized cavity length of 1.7 cm, they achieved laser emission with a threshold pump power of 57 mW and a maximum output power of 0.32 mW at 75 mW pump power [77].

H. Matsuura et al. developed a low-noise yellow-green laser for fluorescence microscopy using a Yb-doped double-clad fiber laser integrated with a periodically poled lithium niobate (PPLN) waveguide crystal. The laser delivered a maximum output of 30 mW with <2% RMS noise across its operating temperature range and an extinction ratio > 22 dB, providing an enhanced light source for fluorescence imaging [78].

Sohel Mahmud Sher et al. proposed the fabrication of embedded channel waveguides in z-cut Er:$LiNbO_3$ substrates via high-energy ion implantation and detailed the theoretical design of continuous-wave (CW) waveguide lasers. Without considering excited-state absorption (ESA), their study revealed an optimal waveguide length of 15–20 mm with a threshold coupled pump power of 25 mW. Introducing ESA significantly reduced output power as the ESA cross-section increased, although the threshold remained unaffected. This work provides critical guidelines for optimizing ion-implanted waveguide lasers [79].

Mathew George et al. reported the first Ti:Tm:$LiNbO_3$ waveguide amplifier operating under 1650 nm in-band pumping. The device exhibited broadband gain spanning 1750–1900 nm, with TE-polarized lasing at 1890 nm achieving a threshold coupled pump power of 4 mW and a slope efficiency of 13.3%. Performance optimization through output mirror reflectivity adjustments highlights its potential for photonic device development [80].

Shilei Ji et al. introduced a novel erbium-doped Ti:$LiNbO_3$ spatial-modulation-gain (SMG) waveguide laser. The device exhibited a lasing threshold at 55 mW pump power, with spontaneous emission noise below threshold and single-frequency operation at 1.5 μm above threshold. At a maximum pump power of 100 mW, this laser demonstrates promising applications in optical communications and LiDAR systems [81].

Yang Tan et al. fabricated a Type II waveguide in Nd:MgO:$LiNbO_3$ crystals using femtosecond laser writing and achieved 1085 nm waveguide lasing. Under 808 nm pumping, the laser demonstrated a slope efficiency of 27%, a threshold of 71 mW, and a maximum output power of 8 mW. These results advance the development of $LiNbO_3$-based integrated photonic devices [82].

B.K. Das et al. reported a novel single-frequency Ti:Er:$LiNbO_3$ distributed Bragg reflector (DBR) waveguide laser, initiating lasing at ~70 mW pump power with an emission wavelength of 1561.84 nm. The device operated in single-longitudinal mode for both TE and TM polarizations, exhibiting a TE-polarized linewidth of 1.11 GHz and delivering 0.72 mW output power at 140 mW pump power. Integrated with two photorefractive gratings as cavity mirrors, this laser shows significant potential for fiber-optic communications and interferometric sensing applications [83].

Jon Martínez de Mendívil et al. fabricated a ridge waveguide laser in Nd-doped $LiNbO_3$ by combining femtosecond laser writing with Zn-diffusion. The laser cavity, formed by bonding multilayer dielectric mirrors to both waveguide ends, achieved TM-polarized lasing at 1085 nm under 815 nm continuous-wave Ti:sapphire pumping. The device demonstrated a threshold of 31 mW and a slope efficiency of 7% [84].

D.B. Rüske et al. demonstrated a high-efficiency Nd-doped Ti-diffused ridge waveguide laser. A 9.5 μm wide, 12 mm long waveguide delivered optimal performance, exhibiting multi-longitudinal-mode operation centered at 1084.7 nm with a full width at a half-maximum (FWHM) of 0.4 nm in Figure 10a. This output validated the laser's wavelength stability and modal characteristics. With only Fresnel end-face reflections (~14%), it achieved a threshold of 83 mW, a maximum output of 108 mW, and a record slope efficiency of 34% for ridge waveguide lasers. Adding high-reflectivity (HR) mirrors reduced the threshold to 59 mW but lowered the slope efficiency to 18%, as shown in Figure 10b, and maximum power to 46 mW. Future work will focus on mitigating photorefractive damage, optimizing output coupling, and refining fabrication parameters to enable single-transverse-mode operation and reduced propagation losses [85].

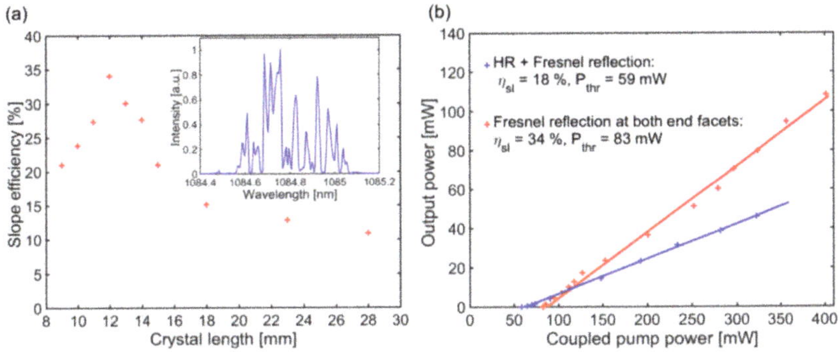

Figure 10. Laser performance of a 9.5 µm wide Nd:Ti:LiNbO$_3$ ridge waveguide. (**a**) Slope efficiency versus crystal length under solely Fresnel-reflection feedback, with an inset showing the output spectrum from the 12 mm long waveguide. (**b**) Output power versus coupled pump power for the 12 mm long waveguide laser comparing Fresnel-only feedback (red line) and feedback with an inserted HR mirror (blue line). Reprinted with permission from [85].

Viacheslav Snigirev et al. reported a rapidly tunable laser based on a hybrid Si$_3$N$_4$/LiNbO$_3$ photonic platform for optical ranging. Figure 11a presents the schematic diagram of the experimental setup, while Figure 11b displays the beat frequency spectra of the targets under different integration times (as indicated by the time slices of 92 ms, 5 ms, and 10 ms in the context of the experiment). The system achieved ~15 cm spatial resolution, detecting target reflections at 2.1 m and wall reflections at 2.8 m in Figure 11c. Processed 3D point-cloud data confirmed its feasibility in LiDAR applications [86].

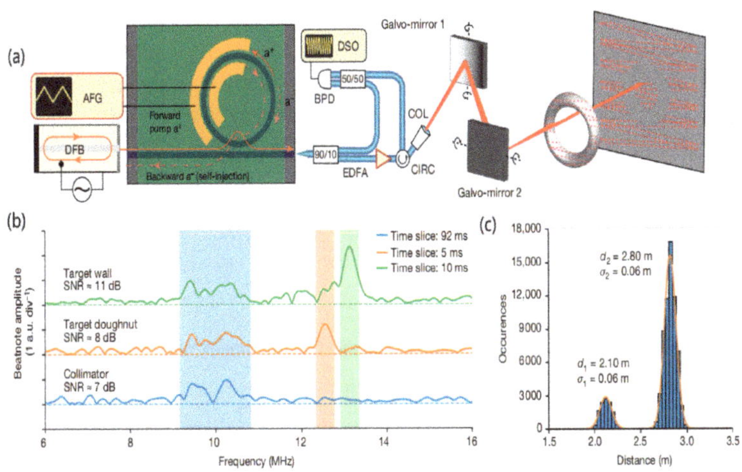

Figure 11. Coherent FMCW LiDAR demonstration using hybrid integrated lasers. (**a**) Experimental schematic for coherent optical ranging based on frequency-modulated continuous-wave (FMCW) LiDAR. (**b**) Representative delayed homodyne beatnote signals showing reflections from the collimator (blue trace, SNR indicated), doughnut target (orange trace, SNR), and wall (green trace, SNR); (**c**) Histogram of measured distance distribution to the target. Reprinted with permission from [86].

Cornelis A. A. Franken et al. reported a high-power, narrow-linewidth thin-film lithium niobate (TFLN) laser fabricated using photonic wire bonding technology. The device exhibited a threshold current of 100 mA, with output power increasing approximately linearly with drive current, achieving on-chip output power up to 76.2 mW, a high side-mode suppression ratio (SMSR) across varying currents, and single frequency operation with 61 dB side mode suppression. Wavelength tuning over 43.7 nm near 1530 nm was achieved by adjusting the effective length of microring resonators in a Vernier filter, as shown in Figure 12. Measured via a delayed self-heterodyne setup, the laser demonstrated an intrinsic linewidth as low as 550 Hz. Remarkably, without active stabilization, it maintained continuous operation for 58 h without mode hopping, with a frequency drift trend of only 4.4 MHz/h [87].

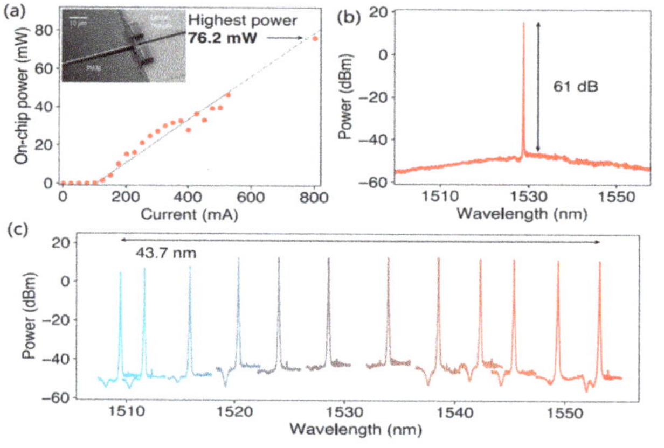

Figure 12. (**a**) Laser power. The inset is a scanning electron microscope image of a photonic wire bond, with an anchoring structure, to a TFLN waveguide. (**b**) Side mode suppression. (**c**) Wave-length tuning. Reprinted with permission from [87].

5. TeO_2

Tellurium dioxide (TeO_2) exhibits significant application potential in laser technology, with its advantageous characteristics primarily stemming from the synergistic effects of multidimensional physical properties. From a material perspective, TeO_2's low phonon energy (700–800 cm^{-1}) effectively suppresses non-radiative transitions, substantially enhancing the radiative transition efficiency of rare-earth ions (e.g., Er^{3+}, Tm^{3+}), making it particularly advantageous for mid-infrared (2–5 μm) laser emission. Studies demonstrate that TeO_2 glass's high refractive index (1.9–2.3), calculated through Judd–Ofelt theory, strengthens electric dipole transition intensity, thereby improving laser gain coefficients and absorption/emission cross-section values. The material system also exhibits exceptional rare-earth ion solubility (e.g., up to 25 mol% Er_2O_3 dissolution), combined with structural diversity (coexistence of TeO_4 bipyramids and TeO_3 trigonal pyramids), which significantly broadens the fluorescence spectral line full width at half maximum (FWHM) (e.g., 60 nm bandwidth for Er^{3+} at 1530 nm), providing a material foundation for broadband tunable lasers and fiber amplifiers. Notably, TeO_2's remarkable nonlinear optical properties (second-order nonlinear refractive index $n_2 \approx 2.5 \times 10^{-19}$ m^2/W) and high thermal stability (glass transition temperature Tg ≈ 280–430 °C) synergistically support high-power laser operation and ultrafast laser applications. These comprehensive characteristics establish TeO_2-based material systems as ideal candidates for near- to mid-infrared laser devices [88].

Khu Vu et al. achieved the first demonstration of an erbium-doped tellurium oxide (TeO_2) planar rib waveguide laser under 980 nm pumping. Random mode-hopping lasing emission, triggered by end-face reflections (~8.5%), was observed at 1532 nm with an output power of 0.1 mW (pump power: 300 mW). By introducing gold-mirror reflections (reflectivity enhanced to >80%) and fiber Bragg grating (FBG) feedback (95% reflectivity), the lasing threshold was reduced to 100 mW, and a narrow-linewidth (<1 GHz) output at 1550.27 nm (as shown in Figure 13a,b) and an output power of 1 mW were achieved. The study revealed significant green upconversion (546 nm emission) under strong pumping but no observable photodarkening effects. Simulations indicate that reducing waveguide losses (<0.2 dB/cm) and optimizing cavity mirror reflectivity could enable lower thresholds (<50 mW) and high-power laser operation, in Er-doped TeO_2 waveguide lasers [89].

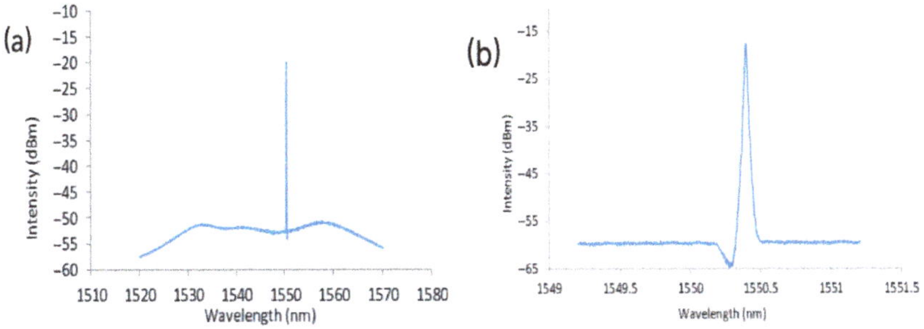

Figure 13. 1550 nm fiber Bragg grating (FBG) stabilized laser emission at 105 mW pumping. (**a**) Wide span spectrum. (**b**) Narrow span spectrum. Reprinted with permission from [89].

Khadijeh et al. successfully demonstrated a compact, monolithically integrated thulium-doped tellurium oxide (TeO_2:Tm^{3+}) microring laser consisting of a 300 µm radius Si_3N_4 microring structure coated with a 0.39-µm-thick TeO_2 gain layer (Tm^{3+} concentration: 4.1×10^{20} cm^{-3}), where 66.7% of the pump energy was confined within the TeO_2 layer. Multimode lasing across 1815–1895 nm with a maximum single-ended output power of 4.5 mW was achieved (as shown in Figure 14), with key performance metrics including a threshold pump power of 18 mW (waveguide-coupled) and 11 mW (microring absorbed), single-ended slope efficiencies of 11% (waveguide-coupled) and 17% (absorbed power), and a bidirectional total efficiency of 34% [90].

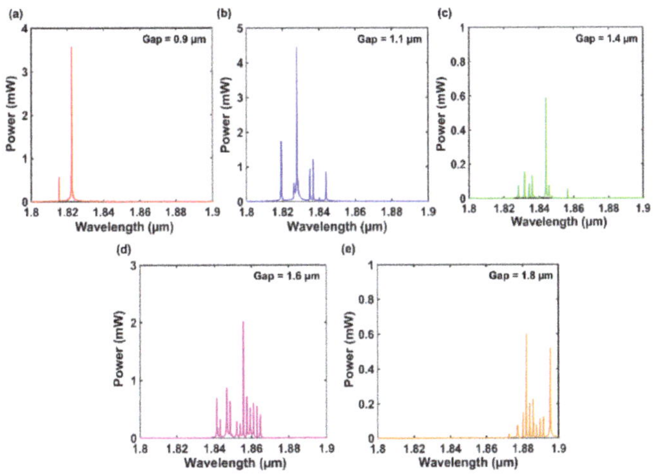

Figure 14. Laser emission spectra generated by 1610 nm optical pumping in TeO_2:Tm^{3+}-Si_3N_4 ring resonators and for microring-waveguide gaps of (**a**) 0.9 µm, (**b**) 1.1 µm, (**c**) 1.4 µm, (**d**) 1.6 µm, and (**e**) 1.8 µm. The laser emission shifts from ~1815 to 1895 nm by increasing the gap size. Reprinted with permission from [90].

Bruno et al. designed an asymmetric grating cavity (input/output grating lengths of 6/3 mm), and demonstrated an integrated on-chip erbium-doped tellurite (TeO_2:Er^{3+}) distributed Bragg reflector (DBR) laser by depositing a 0.352-μm-thick TeO_2:Er^{3+} active layer onto 1.2/1.6 μm wide, 22-mm-long SiN waveguide, where C-band (1530–1565 nm) lasing emission was achieved. The device exhibited a threshold power of 13 mW, a maximum slope efficiency of 0.36% (forward: 0.33%), and a peak output power of 0.28 mW (dual-end pumping). The further improvement of the performance was limited by the background propagation loss (1.0–1.1 dB/cm) and low Er^{3+} ion activation ratios [91].

6. Ta_2O_5

Tantalum pentoxide has a high refractive index (approximately 2.1) that can effectively improve the optical mode constraint capabilities of optical waveguides. Since 2005, a number of studies have successively explored the design and performance optimization of rare earth (Nd^{3+}, Er^{3+}, Yb^{3+}, and Tm^{3+})-doped Ta_2O_5 waveguide lasers. In recent years, advancements in nanofabrication technologies have further extended the Ta_2O_5 platform to applications such as microcavity lasers, waveguide amplifiers, and Kerr nonlinear photonic devices, while heterogeneous integration has enhanced its compatibility with other photonic materials and expanded its functional capabilities.

Unal et al. reported a neodymium-doped tantalum pentoxide (Nd:Ta_2O_5) channel waveguide laser with emissions at 1.066 μm and 1.375 μm. The waveguide loss was 0.2dB/cm. Lasing was achieved at both 1.066 μm and 1.375 μm with a threshold pump power as low as 2.7 mW, and a slope efficiency of 21% was measured [92].

In 2010, Subramanian et al. reported an erbium-doped tantalum pentoxide (Er: Ta_2O_5) bar waveguide laser prepared on silicon oxide using magnetron sputtering. The waveguide is 2.3 cm long, and laser output is realized in the 1558–1562 nm wavelength range under 977 nm laser diode pumping. The launched pump power threshold and slope efficiency were measured to be 14 mW and 0.3%, respectively. The threshold and slope efficiency with respect to absorbed pump power are estimated to be 4.3 mW and 1%, respectively [93].

In 2015, Aghajani et al. reported a Yb:Ta_2O_5 waveguide laser. When pumped with a 977 nm laser diode, laser emission was achieved in the range of 1015–1030 nm. For a cavity formed by a high reflector and about 12% Fresnel reflection at the output end, a maximum output power of 25 mW was obtained at a 1025 nm wavelength, with an absorbed pump power threshold of about 29 mW and a slope efficiency of about 27% [94].

In 2020, Tong et al. reported the spectroscopic characteristics and lasing performance of thulium-doped tantalum pentoxide (Tm:Ta_2O_5) waveguides. They used a 3 μm wide, 1 cm long Tm:Ta_2O_5 rib waveguide, which was end-fire-pumped at 795 nm. At an incident pump power of 170 mW—equivalent to a launched pump power of approximately 44 mW, calculated based on the overlap between the modal field and the incident beam-distinct emission—peaks appeared at 1238 nm and 1855/1858 nm, corresponding to the $^3H_5 \rightarrow {}^3H_6$ and $^3F_4 \rightarrow {}^3H_6$ transitions, respectively. The threshold launched pump power for lasing at 1858 nm was estimated to be around 15 mW, and a gain of at least 9 dB/cm was achieved [95].

In 2021, Jung et al. reported a new platform for integrated nonlinear photonics based on tantalum pentoxide (Ta_2O_5; hereafter, tantala). Tantala films with thicknesses exceeding 800 nm were deposited on oxidized silicon wafers using ion beam sputtering. These films exhibit a low residual tensile stress of 38 MPa and possess a Kerr nonlinear index of $n_2 = 6.2 \pm 2.3 \times 10^{-19}$ m^2/W, which is approximately three times greater than that of silicon nitride. Resonators fabricated from these films achieve optical quality factors as high as 3.8×10^6, enabling the generation of ultrabroadband Kerr soliton frequency combs with low threshold power. Figure 15 illustrates soliton spectra and temporal properties in two RW geometries of tantala resonators, demonstrating efficient outcoupling, reduced thermo-optic effects, and sub-35 fs pulse durations. Leveraging this platform, low-threshold Kerr soliton combs and supercontinuum generation were demonstrated with coupled pulse energies as low as 60 pJ. The nonlinear performance of tantala waveguides was further evaluated through supercontinuum generation using low-energy ultrafast seed pulses, underscoring the potential of tantala for future applications in integrated nonlinear photonics [96].

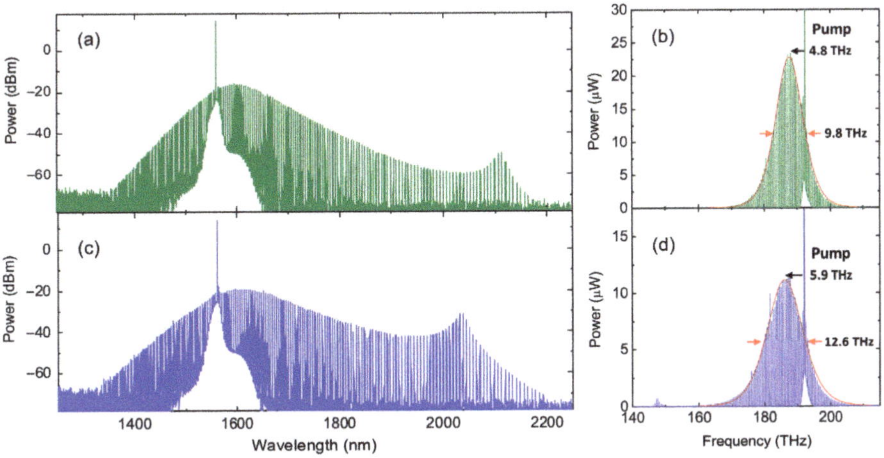

Figure 15. Soliton generation in tantala ring resonators. (**a**) Single-soliton comb spectrum at 1.6 μm pump. (**b**) Linear scale of (**a**), showing FWHM bandwidth. (**c**) Single-soliton comb spectrum at 1.65 μm pump. (**d**) Linear scale of (**c**), highlighting bandwidth. Reprinted with permission from [96].

In 2025, Jian Liu et al. reported the realization of on-chip microdisk lasers and waveguide amplifiers based on erbium-doped tantalum pentoxide (Er:Ta_2O_5). Er:Ta_2O_5 thin films with a thickness of approximately 700 nm were deposited via radio frequency (RF) magnetron sputtering. Subsequent fabrication of photonic structures was carried out using femtosecond direct laser writing in conjunction with chemical mechanical polishing. A suspended Er:Ta_2O_5 microdisk laser with a diameter of 36 μm demonstrated multi-mode lasing under 980 nm optical excitation, achieving a low threshold of 225 μW and exhibiting an intrinsic quality factor of 1.89×10^6 at 951 nm. These results underscore the viability of Er:Ta_2O_5 thin films for the development of high-performance active photonic devices, offering a cost-effective and CMOS-compatible platform for integrated photonic applications [97].

In 2025, a research team reported a heterogeneous integration platform combining InGaAs/GaAs quantum well gain materials with tantala (Ta_2O_5) waveguides, enabling high-yield fabrication of both active and passive photonic integrated circuits (PICs). Distributed feedback (DFB) lasers operating at 976 nm exhibited robust single-mode emission, characterized by a side-mode suppression ratio (SMSR) of 43 dB and a white noise floor in the frequency noise power spectral density (PSD) of approximately 4×10^5 Hz^2/Hz, corresponding to a fundamental linewidth of 1.2 MHz. Integrated semiconductor optical amplifiers (SOAs) achieved a maximum on-chip gain of 24.5 dB at 987.4 nm, with a 3 dB bandwidth of 8.5 nm. Utilizing the DFB lasers to pump high-Q microring resonators (Q $\approx 2.5 \times 10^6$) fabricated in Ta_2O_5, the team demonstrated wide-span degenerate optical parametric oscillation (OPO) over 183 THz, generating short-wavelength signal outputs at 778 nm and 752 nm. Figure 16 showcases a heterogeneous III-V/tantala PIC integrating diverse passive and active components for advanced photonic functionality. This study underscores the promise of Ta_2O_5-based heterogeneous PICs as a scalable and versatile platform for realizing integrated lasers, amplifiers, and nonlinear photonic devices, thereby advancing the development of quantum photonics and compact photonic systems [98].

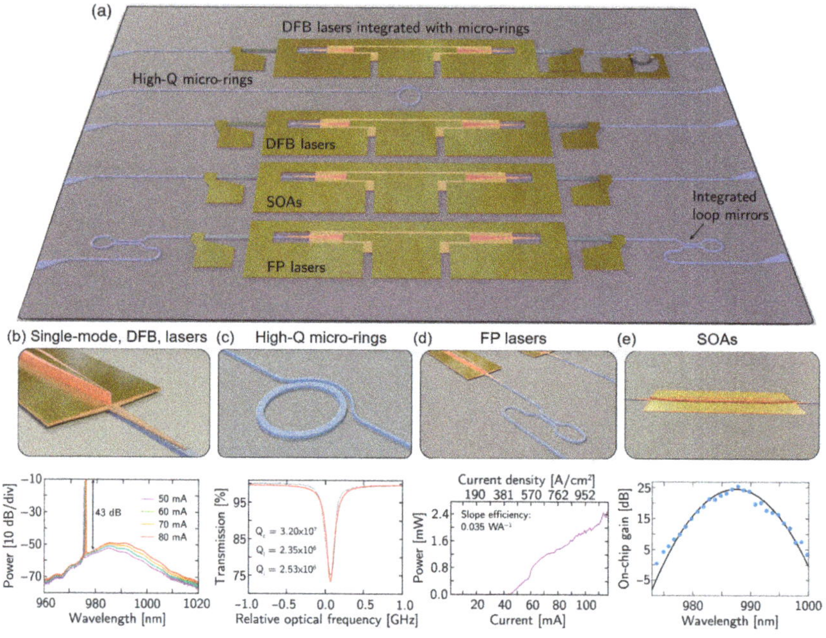

Figure 16. (**a**) 3D view of a heterogeneous tantala PIC enabling wafer-scale integration of active and passive components. (**b**) Single-mode DFB lasers. (**c**) Ultra-low-loss tantala waveguides and high-Q microrings. (**d**) FP lasers with integrated loop mirrors in tantala. (**e**) Tantala-integrated SOA with 24.5 dB gain at 987.4 nm. Reprinted with permission from [98].

7. Outlook and Summary

On-chip integrated lasers are the core of photonic integrated circuits (PICs). They are widely used in high-speed optical communications, high-precision sensing, quantum information processing, and biomedical detection. Their miniaturization, low power consumption, and high stability provide key support for the next generation of optoelectronic systems.

Here we summarize and compare the different on-chip lasers from two aspects:

1. Materials

The gain media cited in the paper can be catalogued into semiconductors and RE-doped gain materials. For the former, since III-V semiconductors are epitaxially grown on silicon wafer, the lattice mismatch between conventional materials (e.g., InP, GaAs) and silicon-based waveguides necessitates complex heterogeneous integration processes with low yields. As a matter of fact, many efforts have been devoted to decreasing the density of various defects in the epitaxial films and increasing the electro-optical conversion efficiency—for example, introducing various buffer layers to reduce the defects or designing the different structures to increase the electro-optical conversion efficiency—but basically, the methods and technologies used in the past or future research for the growth of epitaxial films could be similar. Further improvement of the performance in the semiconductor-based on-chip lasers mostly relies on how to get high-quality semiconductor thin films.

On the other hand, RE-doped materials represent another active research direction in the development of on-chip lasers. One of the key factors for various host materials of RE doping is the low optical loss. Generally, loss less than one-tenth of dB/cm is necessary. In terms of this, SiN and $LiNbO_3$ with low optical losses could be good options, but RE doping into SiN requires ion-plantation, while single crystalline Er:$LiNbO_3$ needs to be bonded into silicon with extra costs. In contrast, various amorphous oxides and nitrides provide opportunities for the development of on-chip lasers, although there are more spaces to reduce the optical loss in these materials. For example, optical loss less than 0.1 dB/cm has been reported in Er:Ta_2O_5 waveguides [98], and thus it is possible to develop high-performance on-chip lasers based on Er:Ta_2O_5. Moreover, we also want to point out that, while RE ions are doped into host materials, host materials with large metallic atomic radius or large bonding length of metal–oxygen are preferred, since this can effectively separate the doped RE ions. In this case, the formation of RE-ion clusters can be avoided, and the emission from the transition between the different energy level of RE ions can be maximized. In terms of this, Ta has a larger atomic radius than Te and Al, and thus, Ta_2O_5 is expected to have better performance in the emission of light. Of course, screening new materials with better RE solubilities, low optical losses, and high gain are always expected for the exciting area.

2. Pump schemes

Comparing different pump schemes, electrically pumped on-chip lasers demonstrate unique advantages. First, the large volume of an internal pump source is replaced by on-chip micro-electrodes, enabling miniaturization of the entire system to the centimeter scale. Second, while optically pumped lasers suffer significant pump energy loss due to coupling inefficiencies between external pump sources and chips, electrically pumped lasers substantially reduce this loss, thereby lowering overall power consumption. Finally, electrically pumped lasers integrate electrodes directly onto the waveguide. In contrast, optically pumped lasers typically require fiber-optic or free-space coupling between the pump laser and chip, making the electrically pumped approach inherently more robust and stable. For practical applications, we believe electrically pumped on-chip lasers are better suited for large-scale production, but this usually causes serious issues of heating management, especially in the case of on-chip lasers with high output power.

Currently, on-chip integrated lasers are facing multiple technical bottlenecks. Regarding fabrication, the lattice mismatch between conventional materials (e.g., InP, GaAs) and silicon-based waveguides necessitates complex heterogeneous integration processes with low yields. In terms of performance, on-chip light sources generally exhibit insufficient output power (typically tens of mW), fundamentally limiting their applicability in long-distance communications or high-sensitivity sensing systems. Concurrently, these devices exhibit low electro-optical conversion efficiency (<30%), inadequate thermal stability, and high power consumption. Additionally, mode–field mismatch results in substantial optical coupling loss (typically 3–5 dB), which severely constrains co-integration efficiency between lasers and passive components.

In response to the above challenges, future research can focus on material innovation and process optimization, including the following:

Heterogeneous integration technology: through wafer bonding and selective epitaxial growth, the efficient integration of Group III-V materials and silicon waveguides can be achieved to reduce the interface defect density.

Mode field matching design: using conical waveguides, metasurface couplers or reverse design algorithms to reduce coupling loss to less than 1 dB.

Thermal management optimization: implementation of an aluminum nitride (AlN) heat dissipation layer or microfluidic cooling structure to improve power density and long-term stability. This is particularly critical for electrically pumped lasers, where even low currents applied to the chip cause significant heat accumulation.

New resonant cavity structure: utilizing distributed feedback (DFB) lasers, a microring resonant cavity, or topological photonic design to enhance light confinement capabilities, reduce threshold current (<1 mA), and improve efficiency. In addition, silicon-based quantum dot lasers and lithium niobate thin-film lasers show higher performance potential through energy band engineering and localization of bright light fields.

A laser light source that is easier to integrate on the chip may be a direct electrically pumped rare-earth-doped material. For example, erbium-doped aluminum nitride (Er:AlN) can achieve 1.55 μm band luminescence through electric field regulation, without complex epitaxial processes, and is compatible with CMOS production lines. By improving the efficiency of rare-earth ions, it is possible to solve the problem of low carrier mobility of oxide materials. It is expected to generate a narrow line–wide laser source from the electric pumped LED source through later mode selection and other devices; another class of gain media receiving significant attention is organic gain media. Organic gain media (such as PMN:PMMA organic dyes and conjugated polymers) can cover spectral ranges from visible to near-infrared light and possess high gain characteristics. Furthermore, solution-based processing methods (like spin coating and inkjet printing) enable localized deposition, offering higher material utilization efficiency. Organic gain media offer a viable pathway for low-cost, integrable visible-light lasers, showing significant potential, especially in silicon photonics. Current challenges require breakthroughs in electrical pumping and stability to achieve commercialization [99]. Additionally, biological protein-based light sources (e.g., fluorescent proteins), through genetic engineering and molecular structure optimization, can enhance output power. These bio-light sources combine biocompatibility with ultra-low power consumption characteristics, making them suitable for applications in implantable medical devices or environmental monitoring [100,101].

On-chip integrated lasers are developing towards high performance, multi-material fusion and interdisciplinary applications. Through breakthroughs in process innovation and new material systems, it is expected to open up new application scenarios in the fields of communications, biology, and quantum information processing.

Author Contributions: Conceptualization, R.W.; methodology, Z.Y., S.B. and R.W.; investigation, Y.Y., J.Y., S.Z. and Z.C.; writing—original draft preparation, Y.Y., J.Y., S.Z., Z.C., S.B. and R.W.; writing—review and editing, Y.Y., J.Y., S.Z., Z.C., Z.Y., S.B. and R.W.; visualization, Y.Y., J.Y., S.Z., Z.C. and S.B.; supervision, R.W.; project administration, R.W.; funding acquisition, Z.Y. and R.W. All authors have read and agreed to the published version of the manuscript.

Funding: The National Natural Science Foundation of China (62475126, 62405354); the 3315 Innovation Team in Ningbo City, Zhejiang Province, China; the K.C. Wong Magna Fund in Ningbo University, China; the Open Project Program of Wuhan National Laboratory for Optoelectronics NO.2022WNLOKF001; and the K.C. Wong Magna Fund in Ningbo University.

Institutional Review Board Statement: Not applicable.

Informed Consent Statement: Not applicable.

Conflicts of Interest: The authors declare no conflicts of interest.

References

1. Shekhar, S.; Bogaerts, W.; Chrostowski, L.; Bowers, J.E.; Hochberg, M.; Soref, R.; Shastri, B.J. Roadmapping the next generation of silicon photonics. *Nat. Commun.* **2024**, *15*, 751. [CrossRef] [PubMed]
2. Zhou, Z.; Chen, W.; He, X.; Ma, D. Photonics in a Time of Rapid Growth: Silicon Based Optoelectronics in China. *IEEE Photonics J.* **2024**, *16*, 0600109. [CrossRef]

3. Zhou, Z.; Chen, W.; Feng, J.; Yang, F.; Ma, D.; He, X.; Li, D.; Hou, H.; Shuai, Y.; Cui, W. Silicon based optoelectronics and its frontier advances. *Acta Opt. Sin.* **2024**, *1*, 21–38. [CrossRef]
4. Hao, Y.; Xiang, S.; Han, G.; Zhang, J.; Ma, X.; Zhu, Z.; Guo, X.; Zhang, Y.; Han, Y.; Song, Z.; et al. Recent progress of integrated circuits and optoelectronic chips. *Sci. China Inf. Sci.* **2021**, *64*, 201401. [CrossRef]
5. Zhang, Z.; Liu, R.; Wang, W.; Yan, K.; Yang, Z.; Song, M.; Wu, D.; Xu, P.; Wang, X.; Wang, R. On-chip Er-Doped Ta_2O_5 waveguide amplifiers with a high internal net gain. *Opt. Lett.* **2023**, *48*, 5799–5802. [CrossRef] [PubMed]
6. Zheng, J.; Fang, Z.; Wu, C.; Zhu, S.; Xu, P.; Doylend, J.K.; Deshmukh, S.; Pop, E.; Dunham, S.; Li, M.; et al. Nonvolatile Electrically Reconfigurable Integrated Photonic Switch Enabled by a Silicon PIN Diode Heater. *Adv. Mater.* **2020**, *32*, 2001218. [CrossRef] [PubMed]
7. Xu, P.; Zheng, J.; Doylend, J.K.; Majumdar, A. Low-Loss and Broadband Nonvolatile Phase-Change Directional Coupler Switches. *ACS Photonics* **2019**, *6*, 553–557. [CrossRef]
8. Feng, D.; Liao, S.; Dong, P.; Feng, N.-N.; Liang, H.; Zheng, D.; Kung, C.-C.; Fong, J.; Shafiiha, R.; Cunningham, J.; et al. High-speed Ge photodetector monolithically integrated with large cross-section silicon-on-insulator waveguide. *Appl. Phys. Lett.* **2009**, *95*, 261105. [CrossRef]
9. Miarabbas Kiani, K.; Frankis, H.C.; Naraine, C.M.; Bonneville, D.B.; Knights, A.P.; Bradley, J.D.B. Lasing in a Hybrid Rare-Earth Silicon Microdisk. *Laser Photonics Rev.* **2021**, *16*, 2100348. [CrossRef]
10. Zhou, Z.; Ou, X.; Fang, Y.; Alkhazraji, E.; Xu, R.; Wan, Y.; Bowers, J.E. Prospects and applications of on-chip lasers. *eLight* **2023**, *3*, 1297–1321. [CrossRef] [PubMed]
11. Shen, H.; Li, D.-S.; Yang, D.-R. Research progress of silicon light source. *Acta Phys. Sin.* **2015**, *64*, 204208-01–204208-18. [CrossRef]
12. Rong, H.; Xu, S.; Kuo, Y.-H.; Sih, V.; Cohen, O.; Raday, O.; Paniccia, M. Low-threshold continuous-wave Raman silicon laser. *Nat. Photonics* **2007**, *1*, 232–237. [CrossRef]
13. Heck, M.J.R.; Bauters, J.F.; Davenport, M.L.; Doylend, J.K.; Jain, S.; Kurczveil, G.; Srinivasan, S.; Tang, Y.; Bowers, J.E. Hybrid Silicon Photonic Integrated Circuit Technology. *IEEE J. Sel. Top. Quantum Electron.* **2013**, *19*, 6100117. [CrossRef]
14. Li, N.; Xin, M.; Su, Z.; Magden, E.S.; Singh, N.; Notaros, J.; Timurdogan, E.; Purnawirman, P.; Bradley, J.D.B.; Watts, M.R. A Silicon Photonic Data Link with a Monolithic Erbium-Doped Laser. *Sci. Rep.* **2020**, *10*, 1114. [CrossRef] [PubMed]
15. Liu, R.; Zhang, Z.; Yang, Z.; Wang, W.; Yan, K.; Song, M.; Wang, R. Erbium-doped Ga_2O_3 waveguide for optical amplification. *Appl. Phys. Lett.* **2023**, *123*, 151109. [CrossRef]
16. Pollnau, M.; Bradley, J.D.B. Optically pumped rare-earth-doped Al_2O_3 distributed-feedback lasers on silicon [Invited]. *Opt. Express* **2018**, *26*, 24164–24189. [CrossRef] [PubMed]
17. Hermans, A.; Van Gasse, K.; Kuyken, B. On-chip optical comb sources. *APL Photonics* **2022**, *7*, 100901. [CrossRef]
18. Chen, P.; Long, Z.; Cheng, Q.; Song, M.; Wang, W.; Liu, R.; Zhang, Z.; Xia, K.; Yang, Z.; Qian, L.; et al. On-Chip Supercontinuum Generation Pumped by Short Wavelength Fiber Lasers. *Photonics* **2024**, *11*, 440. [CrossRef]
19. Ferrara, M.A.; Sirleto, L. Integrated Raman Laser: A Review of the Last Two Decades. *Micromachines* **2020**, *11*, 330. [CrossRef] [PubMed]
20. Tournié, E.; Monge Bartolome, L.; Rio Calvo, M.; Loghmari, Z.; Díaz-Thomas, D.A.; Teissier, R.; Baranov, A.N.; Cerutti, L.; Rodriguez, J.-B. Mid-infrared III–V semiconductor lasers epitaxially grown on Si substrates. *Light Sci. Appl.* **2022**, *11*, 165. [CrossRef] [PubMed]
21. Pan, S.; Cao, V.; Liao, M.; Lu, Y.; Liu, Z.; Tang, M.; Chen, S.; Seeds, A.; Liu, H. Recent progress in epitaxial growth of III–V quantum-dot lasers on silicon substrate. *J. Semicond.* **2019**, *40*, 101302. [CrossRef]
22. Li, N.; Chen, G.; Ng, D.K.T.; Lim, L.W.; Xue, J.; Ho, C.P.; Fu, Y.H.; Lee, L.Y.T. Integrated Lasers on Silicon at Communication Wavelength: A Progress Review. *Adv. Opt. Mater.* **2022**, *10*, 2201008. [CrossRef]
23. Cerutti, L.; Rodriguez, J.B.; Tournie, E. GaSb-Based Laser, Monolithically Grown on Silicon Substrate, Emitting at 1.55 um at Room Temperature. *IEEE Photonics Technol. Lett.* **2010**, *22*, 553–555. [CrossRef]
24. Saxena, D.; Mokkapati, S.; Parkinson, P.; Jiang, N.; Gao, Q.; Tan, H.H.; Jagadish, C. Optically pumped room-temperature GaAs nanowire lasers. *Nat. Photonics* **2013**, *7*, 963–968. [CrossRef]
25. Wang, J.; Ren, X.; Deng, C.; Hu, H.; He, Y.; Cheng, Z.; Ma, H.; Wang, Q.; Huang, Y.; Duan, X.; et al. Extremely Low-Threshold Current Density InGaAs/AlGaAs Quantum-Well Lasers on Silicon. *J. Light. Technol.* **2015**, *33*, 3163–3169. [CrossRef]
26. Spott, A.; Peters, J.; Davenport, M.L.; Stanton, E.J.; Merritt, C.D.; Bewley, W.W.; Vurgaftman, I.; Kim, C.S.; Meyer, J.R.; Kirch, J.; et al. Quantum cascade laser on silicon. *Optica* **2016**, *3*, 545–551. [CrossRef]
27. Feng, M.; Wang, J.; Zhou, R.; Sun, Q.; Gao, H.; Zhou, Y.; Liu, J.; Huang, Y.; Zhang, S.; Ikeda, M.; et al. On-Chip Integration of GaN-Based Laser, Modulator, and Photodetector Grown on Si. *IEEE J. Sel. Top. Quantum Electron.* **2018**, *24*, 8200305. [CrossRef]
28. Feng, M.; He, J.; Sun, Q.; Gao, H.; Li, Z.; Zhou, Y.; Liu, J.; Zhang, S.; Li, D.; Zhang, L.; et al. Room-temperature electrically pumped InGaN-based microdisk laser grown on Si. *Opt. Express* **2018**, *26*, 5043–5051. [CrossRef] [PubMed]
29. Nguyen-Van, H.; Baranov, A.N.; Loghmari, Z.; Cerutti, L.; Rodriguez, J.-B.; Tournet, J.; Narcy, G.; Boissier, G.; Patriarche, G.; Bahriz, M.; et al. Quantum cascade lasers grown on silicon. *Sci. Rep.* **2018**, *8*, 7206. [CrossRef] [PubMed]

30. Loghmari, Z.; Rodriguez, J.B.; Baranov, A.N.; Rio-Calvo, M.; Cerutti, L.; Meguekam, A.; Bahriz, M.; Teissier, R.; Tournié, E. InAs-based quantum cascade lasers grown on on-axis (001) silicon substrate. *APL Photonics* **2020**, *5*, 041302. [CrossRef]
31. Lee, A.D.; Qi, J.; Mingchu, T.; Yunyan, Z.; Seeds, A.J.; Huiyun, L. InAs/GaAs Quantum-Dot Lasers Monolithically Grown on Si, Ge, and Ge-on-Si Substrates. *IEEE J. Sel. Top. Quantum Electron.* **2013**, *19*, 1901107. [CrossRef]
32. Chen, S.; Li, W.; Wu, J.; Jiang, Q.; Tang, M.; Shutts, S.; Elliott, S.N.; Sobiesierski, A.; Seeds, A.J.; Ross, I.; et al. Electrically pumped continuous-wave III–V quantum dot lasers on silicon. *Nat. Photonics* **2016**, *10*, 307–311. [CrossRef]
33. Chen, S.; Liao, M.; Tang, M.; Wu, J.; Martin, M.; Baron, T.; Seeds, A.; Liu, H. Electrically pumped continuous-wave 13 μm InAs/GaAs quantum dot lasers monolithically grown on on-axis Si (001) substrates. *Opt. Express* **2017**, *25*, 4632–4639. [CrossRef] [PubMed]
34. Liu, A.Y.; Peters, J.; Huang, X.; Jung, D.; Norman, J.; Lee, M.L.; Gossard, A.C.; Bowers, J.E. Electrically pumped continuous-wave 13 μm quantum-dot lasers epitaxially grown on on-axis (001) GaP/Si. *Opt. Lett.* **2017**, *42*, 338–341. [CrossRef] [PubMed]
35. Shang, C.; Feng, K.; Hughes, E.T.; Clark, A.; Debnath, M.; Koscica, R.; Leake, G.; Herman, J.; Harame, D.; Ludewig, P.; et al. Electrically pumped quantum-dot lasers grown on 300 mm patterned Si photonic wafers. *Light Sci. Appl.* **2022**, *11*, 299. [CrossRef] [PubMed]
36. Matsuo, S.; Fujii, T.; Hasebe, K.; Takeda, K.; Sato, T.; Kakitsuka, T. Directly Modulated DFB Laser on SiO_2 /Si Substrate for Datacenter Networks. *J. Light. Technol.* **2015**, *33*, 1217–1222. [CrossRef]
37. Wang, Z.; Tian, B.; Pantouvaki, M.; Guo, W.; Absil, P.; Van Campenhout, J.; Merckling, C.; Van Thourhout, D. Room-temperature InP distributed feedback laser array directly grown on silicon. *Nat. Photonics* **2015**, *9*, 837–842. [CrossRef]
38. Wang, J.; Li, X.; Guo, X.; Loh, T.-H.; Ranno, L.; Liu, C.; Rusli; Wang, H.; Brian Sia, J.X. Scalable single-microring hybrid III-V/Si lasers for emerging narrow-linewidth applications. *Opt. Express* **2024**, *32*, 26751–26762. [CrossRef] [PubMed]
39. Duprez, H.; Descos, A.; Ferrotti, T.; Sciancalepore, C.; Jany, C.; Hassan, K.; Seassal, C.; Menezo, S.; Ben Bakir, B. 1310 nm hybrid InP/InGaAsP on silicon distributed feedback laser with high side-mode suppression ratio. *Opt. Express* **2015**, *23*, 8489–8497. [CrossRef] [PubMed]
40. Abbasi, A.; Verbist, J.; Van Kerrebrouck, J.; Lelarge, F.; Duan, G.-H.; Yin, X.; Bauwelinck, J.; Roelkens, G.; Morthier, G. 28 Gb/s direct modulation heterogeneously integrated C-band InP/SOI DFB laser. *Opt. Express* **2015**, *23*, 26479–26485. [CrossRef] [PubMed]
41. Abbasi, A.; Spatharakis, C.; Kanakis, G.; Andre, N.S.; Louchet, H.; Katumba, A.; Verbist, J.; Avramopoulos, H.; Bienstman, P.; Yin, X.; et al. High Speed Direct Modulation of a Heterogeneously Integrated InP/SOI DFB Laser. *J. Light. Technol.* **2016**, *34*, 1683–1687. [CrossRef]
42. Crosnier, G.; Sanchez, D.; Bouchoule, S.; Monnier, P.; Beaudoin, G.; Sagnes, I.; Raj, R.; Raineri, F. Hybrid indium phosphide-on-silicon nanolaser diode. *Nat. Photonics* **2017**, *11*, 297–300. [CrossRef]
43. Zhang, J.; Sun, C.; Xiong, B.; Wang, J.; Hao, Z.; Wang, L.; Han, Y.; Li, H.; Luo, Y.; Xiao, Y.; et al. An InP-based vortex beam emitter with monolithically integrated laser. *Nat. Commun.* **2018**, *9*, 2652. [CrossRef] [PubMed]
44. Tran, M.A.; Huang, D.; Guo, J.; Komljenovic, T.; Morton, P.A.; Bowers, J.E. Ring-Resonator Based Widely-Tunable Narrow-Linewidth Si/InP Integrated Lasers. *IEEE J. Sel. Top. Quantum Electron.* **2020**, *26*, 1500514. [CrossRef]
45. Wong, W.W.; Su, Z.; Wang, N.; Jagadish, C.; Tan, H.H. Epitaxially Grown InP Micro-Ring Lasers. *Nano Lett.* **2021**, *21*, 5681–5688. [CrossRef] [PubMed]
46. Xiang, C.; Guo, J.; Jin, W.; Wu, L.; Peters, J.; Xie, W.; Chang, L.; Shen, B.; Wang, H.; Yang, Q.-F.; et al. High-performance lasers for fully integrated silicon nitride photonics. *Nat. Commun.* **2021**, *12*, 6650. [CrossRef] [PubMed]
47. Kikuchi, T.; Hiratani, T.; Fujiwara, N.; Inoue, N.; Nitta, T.; Eissa, M.; Mitarai, T.; Wang, Y.; Oiso, Y.; Nishiyama, N.; et al. III–V gain region/Si waveguide hybrid lasers with InP-based two-storied ridge structure by direct bonding technology. *Jpn. J. Appl. Phys.* **2022**, *61*, 052002. [CrossRef]
48. Hendriks, W.A.P.M.; Chang, L.; van Emmerik, C.I.; Mu, J.; de Goede, M.; Dijkstra, M.; Garcia-Blanco, S.M. Rare-earth ion doped Al_2O_3 for active integrated photonics. *Adv. Phys. X* **2020**, *6*, 1833753. [CrossRef]
49. van den Hoven, G.N.; Koper, R.J.I.M.; Polman, A.; van Dam, C.; van Uffelen, J.W.M.; Smit, M.K. Net optical gain at 1.53 μm in Er-doped Al_2O_3 waveguides on silicon. *Appl. Phys. Lett.* **1996**, *68*, 1886–1888. [CrossRef]
50. Bradley, J.; Stoffer, R.; Agazzi, L.; Ay, F.; Wörhoff, K.; Pollnau, M. Integrated Al_2O_3: Er^{3+} ring lasers on silicon with wide wavelength selectivity. *Opt. Lett.* **2009**, *35*, 73–75. [CrossRef] [PubMed]
51. Bradley, J.D.B.; Hosseini, E.S.; Purnawirman; Su, Z.; Adam, T.N.; Leake, G.; Coolbaugh, D.; Watts, M.R. Monolithic erbium- and ytterbium-doped microring lasers on silicon chips. *Opt. Express* **2014**, *22*, 12226–12237. [CrossRef] [PubMed]
52. Wang, Y.; Holguín-Lerma, J.A.; Vezzoli, M.; Guo, Y.; Tang, H.X. Photonic-circuit-integrated titanium:sapphire laser. *Nat. Photonics* **2023**, *17*, 338–345. [CrossRef]
53. Yang, J.; Van Gasse, K.; Lukin, D.M.; Guidry, M.A.; Ahn, G.H.; White, A.D.; Vučković, J. Titanium:sapphire-on-insulator integrated lasers and amplifiers. *Nature* **2024**, *630*, 853–859. [CrossRef] [PubMed]

54. Purnawirman; Sun, J.; Adam, T.N.; Leake, G.; Coolbaugh, D.; Bradley, J.D.B.; Hosseini, E.S.; Watts, M.R. C- and L-band erbium-doped waveguide lasers with wafer-scale silicon nitride cavities. *Opt. Lett.* **2013**, *38*, 1760–1762. [CrossRef] [PubMed]
55. Belt, M.; Huffman, T.; Davenport, M.L.; Li, W.; Barton, J.S.; Blumenthal, D.J. Arrayed narrow linewidth erbium-doped waveguide-distributed feedback lasers on an ultra-low-loss silicon-nitride platform. *Opt. Lett.* **2013**, *38*, 4825–4828. [CrossRef] [PubMed]
56. Belt, M.; Blumenthal, D.J. Erbium-doped waveguide DBR and DFB laser arrays integrated within an ultra-low-loss Si_3N_4 platform. *Opt. Express* **2014**, *22*, 10655–10660. [CrossRef] [PubMed]
57. Hosseini, E.S.; Baldycheva, A.; Sun, J.; Bradley, J.; Adam, T.; Leake, G.; Coolbaugh, D.; Watts, M. Erbium-doped laser with multi-segmented silicon nitride structure. In Proceedings of the Optical Fiber Communication Conference, San Francisco, CA, USA, 9–13 March 2014. [CrossRef]
58. Purnawirman; Li, N.; Magden, E.S.; Singh, G.; Singh, N.; Baldycheva, A.; Hosseini, E.S.; Sun, J.; Moresco, M.; Adam, T.N.; et al. Ultra-narrow-linewidth Al_2O_3:Er^{3+} lasers with a wavelength-insensitive waveguide design on a wafer-scale silicon nitride platform. *Opt. Express* **2017**, *25*, 13705–13713. [CrossRef] [PubMed]
59. Su, Z.; Li, N.; Salih Magden, E.; Byrd, M.; Purnawirman; Adam, T.N.; Leake, G.; Coolbaugh, D.; Bradley, J.D.B.; Watts, M.R. Ultra-compact and low-threshold thulium microcavity laser monolithically integrated on silicon. *Opt. Lett.* **2016**, *41*, 5708–5711. [CrossRef] [PubMed]
60. Singh, G.; Purnawirman; Bradley, J.D.B.; Li, N.; Magden, E.S.; Moresco, M.; Adam, T.N.; Leake, G.; Coolbaugh, D.; Watts, M.R. Resonant pumped erbium-doped waveguide lasers using distributed Bragg reflector cavities. *Opt. Lett.* **2016**, *41*, 1189–1192. [CrossRef] [PubMed]
61. Purnawirman; Li, N.; Magden, E.S.; Singh, G.; Moresco, M.; Adam, T.N.; Leake, G.; Coolbaugh, D.; Bradley, J.D.B.; Watts, M.R. Wavelength division multiplexed light source monolithically integrated on a silicon photonics platform. *Opt. Lett.* **2017**, *42*, 1772–1775. [CrossRef] [PubMed]
62. Magden, E.S.; Li, N.; Purnawirman; Bradley, J.D.B.; Singh, N.; Ruocco, A.; Petrich, G.S.; Leake, G.; Coolbaugh, D.D.; Ippen, E.P.; et al. Monolithically-integrated distributed feedback laser compatible with CMOS processing. *Opt. Express* **2017**, *25*, 18058–18065. [CrossRef] [PubMed]
63. Li, N.; Singh, G.; Magden, E.S.; Su, Z.; Singh, N.; Moresco, M.; Leake, G.; Bradley, J.D.B.; Watts, M.R. Reliable Integrated Photonic Light Sources Using Curved Al_2O_3:Er^{3+} Distributed Feedback Lasers. *IEEE Photonics J.* **2017**, *9*, 1504708. [CrossRef]
64. Notaros, J.; Watts, M.R.; Li, N.; Poulton, C.V.; Su, Z.; Byrd, M.J.; Magden, E.S.; Timurdogan, E.; Baiocco, C.; Fahrenkopf, N.M. CMOS-Compatible Optical Phased Array Powered by a Monolithically-Integrated Erbium Laser. *J. Light. Technol.* **2019**, *37*, 5982–5987. [CrossRef]
65. Li, N.; Purnawirman; Su, Z.; Salih Magden, E.; Callahan, P.T.; Shtyrkova, K.; Xin, M.; Ruocco, A.; Baiocco, C.; Ippen, E.P.; et al. High-power thulium lasers on a silicon photonics platform. *Opt. Lett.* **2017**, *42*, 1181–1184. [CrossRef] [PubMed]
66. Li, N.; Magden, E.S.; Su, Z.; Singh, N.; Ruocco, A.; Xin, M.; Byrd, M.; Callahan, P.T.; Bradley, J.D.B.; Baiocco, C.; et al. Broadband 2-μm emission on silicon chips: Monolithically integrated Holmium lasers. *Opt. Express* **2018**, *26*, 2220–2230. [CrossRef] [PubMed]
67. Becker, P.; Brinkmann, R.; Dinand, M.; Sohler, W.; Suche, H. Er-diffused Ti:$LiNbO_3$ waveguide laser of 1563 and 1576 nm emission wavelengths. *Appl. Phys. Lett.* **1992**, *61*, 1257–1259. [CrossRef]
68. Amin, J.; Hempstead, M.; Roman, J.E.; Wilkinson, J.S. Tunable coupled-cavity waveguide laser at room temperature in Nd-diffused Ti:$LiNbO_3$. *Opt. Lett.* **1994**, *19*, 1541–1543. [CrossRef] [PubMed]
69. Amin, J.; Aust, J.A.; Sanford, N.A. Z-propagating waveguide lasers in rare-earth-doped Ti:$LiNbO_3$. *Appl. Phys. Lett.* **1996**, *69*, 3785–3787. [CrossRef]
70. Jones, J.; De Sandro, J.; Hempstead, M.; Shepherd, D.; Large, A.; Tropper, A.; Wilkinson, J. Channel waveguide laser at 1 μm in Yb-indiffused $LiNbO_3$. *Phys. Rev. Lett.* **1995**, *20*, 1477–1479. [CrossRef]
71. Baumann, I.; Brinkmann, R.; Dinand, M.; Sohler, W.; Westenhofer, S. Ti: Er: $LiNbO_3$ waveguide laser of optimized efficiency. *IEEE J. Quantum Electron.* **1996**, *32*, 1695–1706. [CrossRef]
72. Balsamo, S.; Maio, S.; Montrosset, I.; Suche, H.; Sohler, W. Q-switched Ti: Er: $LiNbO_3$ waveguide laser. *Opt. Quantum Electron.* **1999**, *31*, 29–33. [CrossRef]
73. Das, B.K.; Suche, H.; Sohler, W. Single-frequency Ti:Er:$LiNbO_3$ distributed Bragg reflector waveguide laser with thermally fixed photorefractive cavity. *Appl. Phys. B* **2001**, *73*, 439–442. [CrossRef]
74. Di Paolo, R.; Cantelar, E.; Pernas, P.; Lifante, G.; Cusso, F. Continuous wave waveguide laser at room temperature in Nd^{3+}-doped Zn: $LiNbO_3$. *Appl. Phys. Lett.* **2001**, *79*, 4088–4090. [CrossRef]
75. Fujimura, M.; Tsuchimoto, H.; Suhara, T. Yb-diffused $LiNbO_3$ annealed/proton-exchanged waveguide lasers. *IEEE Photonics Technol. Lett.* **2005**, *17*, 130–132. [CrossRef]
76. Fujimura, M.; Tsuchimoto, H.; Suhara, T. Yb:$LiNbO_3$ Annealed/Proton-Exchanged Waveguide Lasers Pumped by InGaAs Laser Diode at 980 nm Wavelength. *Jpn. J. Appl. Phys.* **2007**, *46*, 5447–5449. [CrossRef]

77. Fujimura, M.; Tsuchimoto, H.; Suhara, T. Yb-indiffused $LiNbO_3$ annealed/proton-exchanged waveguide lasers. *Electron. Commun. Jpn.* **2007**, *90*, 10–17. [CrossRef]
78. Matsuura, H.; Takagi, T. Development of a Low-Noise Yellow-Green Laser Using a Yb-Doped Double-Clad Fiber Laser and a Periodically Poled $LiNbO_3$ Waveguide Crystal. *Jpn. J. Appl. Phys.* **2011**, *50*, 032701. [CrossRef]
79. Sher, S.M.; Pintus, P.; Di Pasquale, F.; Bianconi, M.; Montanari, G.B.; De Nicola, P.; Sugliani, S.; Prati, G. Design of 980 nm-Pumped Waveguide Laser for Continuous Wave Operation in Ion Implanted Er:$LiNbO_3$. *IEEE J. Quantum Electron.* **2011**, *47*, 526–533. [CrossRef]
80. George, M.; Ricken, R.; Quiring, V.; Sohler, W. In-band pumped Ti:Tm:$LiNbO_3$ waveguide amplifier and low threshold laser. *Laser Photonics Rev.* **2012**, *7*, 122–131. [CrossRef]
81. Jin, S.; Xu, L.; Li, Y. Spatially Modulated Gain Erbium-Doped Ti:$LiNbO_3$ Waveguide Laser. *IEEE Photonics Technol. Lett.* **2014**, *26*, 1515–1517. [CrossRef]
82. Tan, Y.; de Aldana, J.R.V.; Chen, F. Femtosecond laser-written lithium niobate waveguide laser operating at 1085 nm. *Opt. Eng.* **2014**, *53*, 107109. [CrossRef]
83. Das, B.; Suche, H.; Sohler, W. Single frequency Ti: Er: $LiNbO_3$ distributed Bragg reflector (DBR)-waveguide laser with two integrated photorefractive gratings. In Proceedings of the European Conference on Integrated Optics, Paderborn, Germany, April 2001; pp. 87–90. [CrossRef]
84. Martínez de Mendívil, J.; del Hoyo, J.; Solís, J.; Lifante, G. Ridge waveguide laser in Nd:$LiNbO_3$ by Zn-diffusion and femtosecond-laser structuring. *Opt. Mater.* **2016**, *62*, 353–356. [CrossRef]
85. Brüske, D.; Suntsov, S.; Rüter, C.E.; Kip, D. Efficient Nd:Ti:LiNbO3 ridge waveguide lasers emitting around 1085 nm. *Opt. Express* **2019**, *27*, 8884–8889. [CrossRef] [PubMed]
86. Snigirev, V.; Riedhauser, A.; Lihachev, G.; Churaev, M.; Riemensberger, J.; Wang, R.N.; Siddharth, A.; Huang, G.; Möhl, C.; Popoff, Y.; et al. Ultrafast tunable lasers using lithium niobate integrated photonics. *Nature* **2023**, *615*, 411–417. [CrossRef] [PubMed]
87. Franken, C.A.A.; Cheng, R.; Powell, K.; Kyriazidis, G.; Rosborough, V.; Musolf, J.; Shah, M.; Barton, D.R.; Hills, G.; Johansson, L.; et al. High-power and narrow-linewidth laser on thin-film lithium niobate enabled by photonic wire bonding. *APL Photonics* **2025**, *10*, 026107. [CrossRef]
88. Jha, A.; Richards, B.; Jose, G.; Teddy-Fernandez, T.; Joshi, P.; Jiang, X.; Lousteau, J. Rare-earth ion doped TeO_2 and GeO_2 glasses as laser materials. *Prog. Mater. Sci.* **2012**, *57*, 1426–1491. [CrossRef]
89. Vu, K.; Farahani, S.; Madden, S. 980 nm pumped erbium doped tellurium oxide planar rib waveguide laser and amplifier with gain in S, C and L band. *Opt. Express* **2015**, *23*, 747–755. [CrossRef] [PubMed]
90. Miarabbas Kiani, K.; Frankis, H.C.; Mateman, R.; Leinse, A.; Knights, A.P.; Bradley, J.D.B. Thulium-doped tellurium oxide microring lasers integrated on a low-loss silicon nitride platform. *Opt. Mater. Express* **2021**, *11*, 3656–3665. [CrossRef]
91. Segat Frare, B.L.; Torab Ahmadi, P.; Hashemi, B.; Bonneville, D.B.; Mbonde, H.M.; Frankis, H.C.; Knights, A.P.; Mascher, P.; Bradley, J.D.B. On-chip hybrid erbium-doped tellurium oxide–silicon nitride distributed Bragg reflector lasers. *Appl. Phys. B* **2023**, *129*, 158. [CrossRef]
92. Unal, B.; Netti, M.C.; Hassan, M.A.; Ayliffe, P.J.; Charlton, M.D.B.; Lahoz, F.; Perney, N.M.B.; Shepherd, D.P.; Chao-Yi, T.; Wilkinson, J.S.; et al. Neodymium-doped tantalum pentoxide waveguide lasers. *IEEE J. Quantum Electron.* **2005**, *41*, 1565–1573. [CrossRef]
93. Subramani, A.Z.; Oton, C.J.; Shepherd, D.P.; Wilkinson, J.S. Erbium-Doped Waveguide Laser in Tantalum Pentoxide. *IEEE Photonics Technol. Lett.* **2010**, *22*, 1571–1573. [CrossRef]
94. Aghajani, A.; Murugan, G.S.; Sessions, N.P.; Apostolopoulos, V.; Wilkinson, J.S. Waveguide lasers in ytterbium-doped tantalum pentoxide on silicon. *Opt. Lett.* **2015**, *40*, 2549–2552. [CrossRef] [PubMed]
95. Tong, A.S.K.; Mitchell, C.J.; Aghajani, A.; Sessions, N.; Senthil Murugan, G.; Mackenzie, J.I.; Wilkinson, J.S. Spectroscopy of thulium-doped tantalum pentoxide waveguides on silicon. *Opt. Mater. Express* **2020**, *10*, 2201–2211. [CrossRef]
96. Jung, H.; Yu, S.-P.; Carlson, D.R.; Drake, T.E.; Briles, T.C.; Papp, S.B. Tantala Kerr nonlinear integrated photonics. *Optica* **2021**, *8*, 811–817. [CrossRef]
97. Liu, J.; Hou, Q.; Zhou, Y.; Dai, Q.; Liu, Z.; Wang, M.; Gao, R.; Fang, Z.; Lin, J.; Cheng, Y. On-chip microlaser and waveguide amplifier on thin film erbium-doped tantalum pentoxide. *Chin. Opt. Lett.* **2025**, *23*, 071302. [CrossRef]
98. Nader, N.; Stanton, E.J.; Brodnik, G.M.; Jahan, N.; Weight, S.C.; Williams, L.M.; Eshaghian Dorche, A.; Silverman, K.L.; Nam, S.W.; Papp, S.B.; et al. Heterogeneous tantala photonic integrated circuits for sub-micron wavelength applications. *Optica* **2025**, *12*, 585–593. [CrossRef]
99. Vogelbacher, F.; Sagmeister, M.; Kraft, J.; Zhou, X.; Huang, J.; Li, M.; Jiang, K.-J.; Song, Y.; Unterrainer, K.; Hainberger, R. Slot-Waveguide Silicon Nitride Organic Hybrid Distributed Feedback Laser. *Sci. Rep.* **2019**, *9*, 18438. [CrossRef] [PubMed]

100. Pris, A.D.; Utturkar, Y.; Surman, C.; Morris, W.G.; Vert, A.; Zalyubovskiy, S.; Deng, T.; Ghiradella, H.T.; Potyrailo, R.A. Towards high-speed imaging of infrared photons with bio-inspired nanoarchitectures. *Nat. Photonics* **2012**, *6*, 195–200. [CrossRef]
101. Gather, M.C.; Yun, S.H. Bio-optimized energy transfer in densely packed fluorescent protein enables near-maximal luminescence and solid-state lasers. *Nat. Commun.* **2014**, *5*, 5722. [CrossRef] [PubMed]

photonics MDPI

Review

Progress in Passive Silicon Photonic Devices: A Review

Qidi Liu [1,*], Yusheng Bian [2] and Jiawei Xiong [3]

1 School of Electrical and Computer Engineering, University of Georgia, Athens, GA 30605, USA
2 Department of Electrical Engineering, The Pennsylvania State University, State College, PA 16802, USA; yushengbian@gmail.com
3 Department of Education Psychology, University of Georgia, Athens, GA 30605, USA; jiawei.xiong@uga.edu
* Correspondence: qidi.liu.photonics@gmail.com

Abstract

Silicon photonics has emerged as a critical enabling technology for a diverse range of applications, from high-speed data communication and computing to advanced sensing and quantum information processing. This paper provides a comprehensive review of recent progress in the foundational passive devices that underpin this technological revolution. We survey the state of the art in fundamental building blocks, including strip, rib, and silicon nitride waveguides, with a focus on achieving ultra-low propagation loss. The review details essential components for light coupling and splitting, such as grating couplers, edge couplers, multimode interference couplers, and directional couplers, citing their typical performance metrics. Key wavelength filtering and routing components, including high-Q ring resonators, Mach–Zehnder interferometers, and arrayed waveguide gratings, are analyzed. Furthermore, we provide a comparative overview of the capabilities of major photonic foundries operating on a multi-project wafer model. The paper concludes by discussing persistent challenges in packaging and polarization management, and explores future trends driven by co-packaged optics, inverse design methodologies, and the expansion of silicon photonics into new application domains.

Keywords: silicon photonics; passive photonic devices; integrated optics

1. Introduction

1.1. The Imperative for Silicon Photonics in the AI and Zettabyte Era

The relentless expansion of the digital economy, propelled by the explosive growth of artificial intelligence (AI) [1], machine learning (ML) [2], cloud computing [3], and high-performance computing (HPC) [4], has created an unprecedented demand for data movement [5]. This surge in traffic, particularly within and between the massive data centers that form the backbone of modern computation, has pushed traditional copper-based electrical interconnects to their fundamental physical limits. As data rates climb, copper links face overwhelming challenges in bandwidth, power consumption, signal integrity, and physical density. This has created a critical bottleneck where data transfer, rather than data processing, becomes the primary constraint on system performance, especially for AI workloads that require massive, parallel data exchange between processors, memory, and storage.

Silicon photonics (SiPh) has emerged as a transformative technology to resolve this interconnect crisis. By using photons to carry information instead of electrons, SiPh leverages the intrinsic advantages of optical communication: much higher bandwidth, lower latency, significantly reduced power consumption per bit, and complete immunity to electromagnetic interference (EMI). In today's "Zettabyte Era", SiPh technology is more important than ever. The performance of AI systems, including the large language models behind generative AI, is directly held back by the hardware's ability to handle data coming in and out. Indeed, over 90% of the inference operations in the neural networks foundational to these models are constrained by data transfer bottlenecks that silicon photonics is uniquely positioned to solve [6].

A pivotal factor accelerating the adoption of SiPh is its profound synergy with the mature Complementary Metal–Oxide–Semiconductor (CMOS) manufacturing ecosystem. This convergence allows for the fabrication of complex photonic integrated circuits (PICs) using the same high-volume, high-precision, and cost-effective processes developed over decades for the microelectronics industry. This manufacturability at scale is enabling a historic transition from discrete optical components to highly integrated systems on a chip, mirroring the evolutionary path of electronic integrated circuits [7]. As the technology matures, it is becoming the foundational platform not only for next-generation data centers and 5G/6G telecommunications but also for disruptive new applications in quantum computing, LiDAR for autonomous systems, and advanced biomedical and environmental sensing [8,9].

1.2. Passive vs. Active Devices in PICs

A photonic integrated circuit is a microchip that contains two or more photonic components to form a functioning circuit, manipulating light on a semiconductor substrate. These components can be broadly categorized into two classes: passive and active.

Passive optical components are devices that perform their function without requiring external power or active control. They are the fundamental pipes of a PIC, responsible for manipulating the flow of light through processes such as guiding, splitting, combining, filtering, and coupling. Examples include waveguides, which confine and direct light; couplers and splitters, which distribute optical power; and filters, which select specific wavelengths, as shown in Figure 1. These components are essential for managing and routing optical signals with minimal loss and interference, forming the foundational infrastructure upon which all other on-chip functionalities are built.

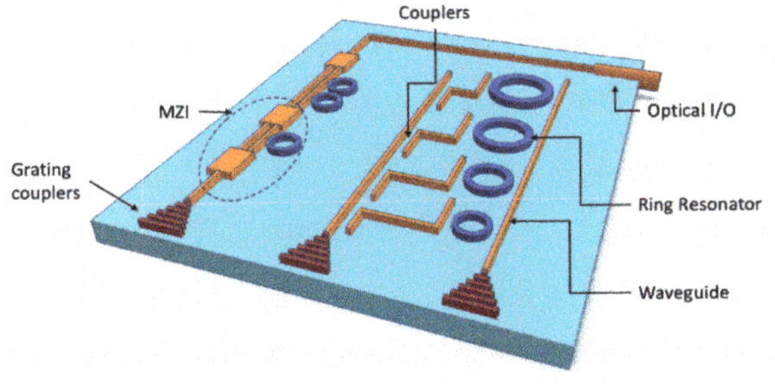

Figure 1. Examples of common silicon photonic passive devices.

In contrast, active devices require an external power source to perform their function, which typically involves converting signals between the electrical and optical domains or actively modifying the properties of the light signal itself. This category includes modulators, which encode electrical data onto an optical carrier; photodetectors, which convert optical signals back into electrical ones; and light sources such as lasers and amplifiers.

This review will focus exclusively on the state of the art and recent progress in passive silicon photonic devices. While active components are critical for a complete system, the performance, efficiency, and scalability of any PIC are ultimately supported by the quality and sophistication of its passive device library.

1.3. Overview of the Review

This paper provides a comprehensive review of the recent progress in passive silicon photonic devices. Section 2 begins with the most fundamental building block, the optical waveguide, reviewing the primary architectures (strip, rib, SiN) and the advanced concept of sub-wavelength grating metamaterials. Section 3 examines the critical components for light coupling and splitting, detailing the performance of fiber-to-chip couplers and on-chip power splitters. Section 4 delves into devices for wavelength filtering and routing, including microring resonators, Mach–Zehnder interferometers, and arrayed waveguide gratings. Section 5 presents a comparative overview of the global fabrication landscape, introducing the multi-project wafer model and profiling the capabilities of major silicon photonics foundries. Section 6 discusses the persistent challenges facing the field, such as polarization management and packaging, and explores future trends including co-packaged optics, inverse design, and new applications in quantum and sensing. Finally, Section 7 provides a summary of the key findings and a concluding perspective on the field's trajectory.

2. Fundamental Building Blocks: Waveguides for Light Confinement and Propagation

The optical waveguide is the most fundamental component in any photonic integrated circuit, serving as the channel for confining and propagating light across the chip. The performance of virtually all other passive and active devices is directly dependent on the properties of the underlying waveguide. The primary goal in waveguide design is to achieve low propagation loss while maintaining sufficient optical confinement to enable compact routing and device footprints. This section reviews the principal waveguide architectures that form the foundation of modern silicon photonics.

2.1. Conventional Silicon Waveguide Architectures on SOI

The silicon-on-insulator (SOI) platform is the most mature and widely used material system for silicon photonics. It consists of a thin crystalline silicon device layer on top of a thicker buried oxide (BOX) layer, which itself sits on a silicon handle wafer. The large refractive index contrast between the silicon core ($n \approx 3.47$ at 1550 nm) and the silicon dioxide (SiO_2) cladding ($n \approx 1.45$) enables strong light confinement [10].

2.2. Strip Waveguides

Strip waveguides, as shown in Figure 2a, are formed by fully etching through the silicon device layer to define a rectangular core, typically with sub-micron cross-sectional dimensions (e.g., 450–500 nm wide and 220 nm high for single-mode operation at 1550 nm) [11].

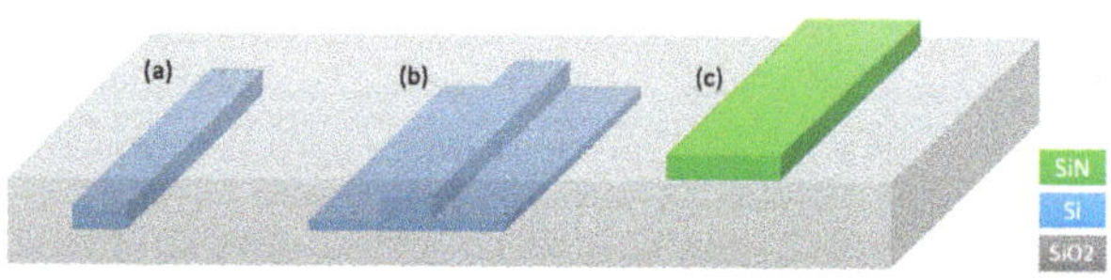

Figure 2. Schematic of (**a**) strip, (**b**) rib, and (**c**) SiN waveguides.

The primary advantage of this architecture is its extremely strong optical confinement. This high confinement allows for very tight bending radii, as small as a few micrometers, without significant radiation loss. This capability is critical for realizing highly compact devices, such as microring resonators with a large free spectral range (FSR), and for enabling dense integration of complex circuits on a single chip.

However, this strong confinement comes at a cost. The optical mode is tightly concentrated within the core, leading to significant interaction with the waveguide's top, bottom, and sidewall interfaces. While the top and bottom surfaces are typically very smooth, the vertical sidewalls are defined by the etching process, which inevitably introduces roughness. This sidewall roughness is a major source of scattering, which is the dominant loss mechanism in strip waveguides. The scattering loss scales strongly with the refractive index contrast, making it particularly pronounced in the high-contrast SOI system.

Historically, propagation losses for strip waveguides were in the range of 1–3 dB/cm [12,13]. However, continuous improvements in lithography and etching processes have led to significant reductions. More recent work on standard 220 nm SOI platforms has demonstrated propagation losses for transverse-electric (TE) polarized light of 2.4 dB/cm [14]. Furthermore, research has shown that operating at longer wavelengths can reduce loss; a record-low propagation loss of approximately 0.2 dB/cm was achieved for a 700 nm wide single-mode strip waveguide operating in the 2 μm wavelength band. This improvement is attributed to the weaker optical field intensity at the waveguide sidewalls at longer wavelengths compared to the standard 1550 nm band [15].

2.3. Rib Waveguides

Rib waveguides are designed to directly address the challenge of sidewall scattering loss. In this architecture, the silicon device layer is only partially etched, leaving a thinner slab of silicon on either side of the central, thicker guiding region [16]. This geometry alters the mode profile, allowing the optical field to spread laterally into the un-etched slab regions [17]. Consequently, the mode's overlap with the rough, vertically etched sidewalls is significantly reduced, leading to a dramatic decrease in propagation loss, as shown in Figure 2b.

The primary advantage of the rib waveguide is its potential for ultra-low loss. The trade-off for this reduced loss is weaker optical confinement compared to a strip waveguide. This weaker confinement necessitates much larger bending radii, typically in the range of tens of micrometers or even millimeters to prevent excessive radiation loss [18]. This limits the ultimate integration density achievable with rib waveguide-based circuits.

The performance of rib waveguides has seen remarkable progress, with reported losses moving from the dB/cm to the dB/m regime, a critical achievement for applications requiring long optical paths like delay lines, gyroscopes, and ultra-high-Q resonators. Early work on shallow-ridge waveguides on 250 nm SOI achieved an average propagation loss of 0.274 dB/cm using standard lithography and dry etching, a figure that was state-of-the-art at the time [11]. A key strategy for further loss reduction is to use thicker SOI wafers and a very shallow etch. By fabricating rib waveguides on a 500 nm thick SOI platform with a rib height of only ~10% of the total silicon thickness (a 56 nm etch), researchers have demonstrated propagation losses as low as 4 dB/m (0.04 dB/cm) [16]. An alternative fabrication approach involves using anisotropic wet etching with chemicals like tetramethylammonium hydroxide (TMAH). This process selectively etches crystal planes, resulting in waveguide sidewalls that are atomically smooth. This technique has been used to achieve propagation losses of 0.85 dB/cm for TE polarization and 1.08 dB/cm for TM polarization [10].

This evolution highlights a fundamental trade-off in waveguide design between optical confinement and propagation loss. While strip waveguides prioritize density and rib waveguides prioritize low loss, advanced fabrication techniques are continuously mitigating this trade-off, enabling complex, large-scale PICs that require both low loss and reasonable compactness.

2.4. The Silicon Nitride (SiN) Platform: A Key Alternative

While SOI is the dominant platform, silicon nitride (SiN) has emerged as a powerful and complementary material for passive photonic devices, as shown in Figure 2c. SiN is a dielectric material that is fully compatible with CMOS fabrication processes and offers a unique set of properties that make it highly attractive for specific application [19]. It has a moderate refractive index of approximately 2.0, which provides good confinement when cladded with SiO_2 while being less sensitive to fabrication imperfections than the higher-contrast silicon system [20].

The key advantages of the SiN platform are its ultra-low propagation loss, broad transparency window, and high-power handling. SiN waveguides are renowned for their exceptionally low propagation losses, which are now routinely in the dB/m range [21]. This is primarily due to the lower refractive index contrast, which reduces scattering from sidewall roughness, and the high quality of SiN films deposited via processes like Low-Pressure Chemical Vapor Deposition (LPCVD). Recent demonstrations have reported losses as low as 0.045 dB/m [22], 1.4 dB/m [23], and, after high-temperature annealing to drive out hydrogen impurities, an astonishing 0.06 dB/cm [24]. Foundries like AIM Photonics now target SiN waveguide losses of below 1 dB/cm in their standard processes [25,26].

Unlike silicon, which becomes opaque below ~1.1 μm, SiN is transparent over a very wide wavelength range, from the visible (~400 nm) to the mid-infrared. This makes it an ideal platform for applications outside the traditional telecom bands, such as visible-light photonics for quantum information and displays, or MIR sensing [27]. Furthermore, SiN exhibits negligible two-photon absorption at telecom wavelengths, a nonlinear effect that can be a limiting factor in silicon waveguides. This property allows SiN devices to handle much higher optical powers, making them the platform of choice for nonlinear applications like frequency comb generation and parametric amplification [19].

A significant challenge in SiN photonics has been the high intrinsic tensile stress of the films, particularly for the thick layers (≥600 nm) required for dispersion engineering in nonlinear applications. This stress can lead to film cracking and wafer bowing, especially on large 300 mm wafers. A breakthrough solution to this problem is the photonic Damascene process [28]. In this method, instead of depositing SiN and then etching it, trenches are first etched into the SiO_2 cladding. SiN is then deposited to fill these trenches, and the excess is removed via chemical mechanical planarization (CMP). This process effectively mitigates the stress issue, enabling the fabrication of ultra-low-loss, thick SiN waveguides on large-scale wafers. As shown in Figure 3, the photonic Damascene process has been demonstrated to enable the high-yield, wafer-scale fabrication of high-performance, dispersion-engineered SiN photonic circuits with ultra-low loss [29]. In conclusion, Table 1 provides a summary of the key performance metrics for each Si and SiN waveguide.

Table 1. Waveguide performance on Si and SiN platform.

Ref	Loss (dB/cm)	Wavelength	Platform	Bend Loss (dB)	Core Thickness	Wafer Size (inch)	Polarization Handling
[24]	0.06	Cband	SIN	NA	730 nm	4	TE
[21]	0.34	Oband	SIN	0.018 for 50 μm	NA	NA	TE
[11]	0.274 ± 0.008	Cband	SOI	$<10^{-4}$ for 100 μm	250 nm	6	TE
[28]	0.055	Cband	SIN	NA	870 nm	NA	TE/TM
[23]	0.014 @1.55 μm	Cband	SIN	NA	720 nm	4	TE

Table 1. *Cont.*

Ref	Loss (dB/cm)	Wavelength	Platform	Bend Loss (dB)	Core Thickness	Wafer Size (inch)	Polarization Handling
[12]	TE: 3.6 ± 0.1 @1.5 µm, TE: 6.8 ± 0.2 @1.3 µm, TM: 3.5 ± 0.1 @1.3 µm	1.2–1.7 µm	SOI	0.086 ± 0.005 for 1 µm, 0.013 ± 0.005 for 2 µm	220 nm	8	TE/TM
[19]	0.157 @1.55 µm, 0.06 @1.58 µm	1.49–1.58 µm	SIN	0.013 for 50 µm, 0.0095 for 80 µm	400 nm	8	TE/TM
[13]	2.01 ± 0.231	Cband	SOI	0.05 ± 0.005 for 30 µm	50 nm	6	TE
[22]	0.00045	1.58–1.61 µm	SIN	NA	50 nm	NA	TE
[16]	0.16	C + L band	SOI	NA	500 nm	NA	TE

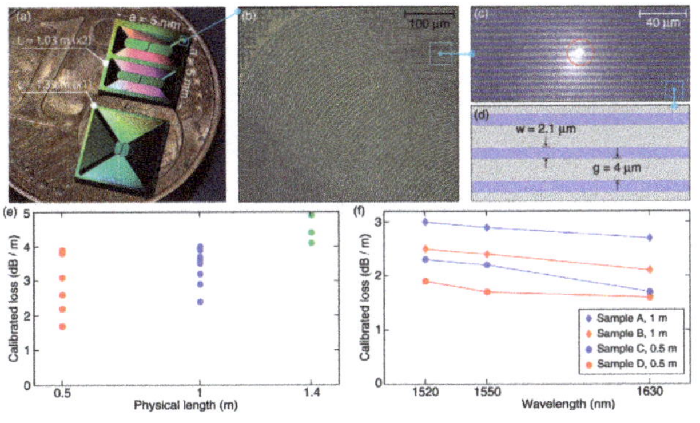

Figure 3. (**a**–**c**): Photos of Si3N4 chips featuring spiral waveguides up to 1.4 m long. An infrared camera is used to highlight light-scattering defects. (**d**): A schematic diagram showing the waveguide width *d* and spacing *g*. (**e**,**f**): Illustrating the measured optical losses at different waveguide lengths *l* and wavelengths. Reprinted with permission from [29].

2.5. Advanced Waveguide Engineering: Sub-Wavelength Grating (SWG) Metamaterials

Sub-wavelength grating (SWG) waveguides represent a paradigm shift in device design, moving from using silicon as a simple wave-guiding medium to engineering its properties at a sub-wavelength scale to create an artificial optical material or metamaterial [30–32]. An SWG waveguide consists of a periodic arrangement of silicon segments interleaved with a low-index material (like the SiO_2 cladding), where the period of the grating, or pitch (Λ), is small enough to suppress diffraction effects. For light propagating along the grating, if the pitch is below the Bragg condition ($\Lambda < \frac{\lambda_0}{2n_{eff}}$), the structure behaves not as a series of discrete scatterers but as a continuous, homogeneous waveguide with unique, engineered properties.

The key characteristic of SWGs is the ability to achieve index engineering. By intentionally designing the geometry of the sub-wavelength segments, the effective refractive index of the equivalent metamaterial can be precisely controlled. The periodic structure is inherently anisotropic, meaning its effective refractive index is different for electric fields oscillating parallel or perpendicular to the grating segments. This engineered anisotropy provides an extra degree of freedom unavailable in conventional solid-core waveguides.

This design flexibility has enabled breakthrough performance and novel functionalities across a wide range of passive devices. SWGs are used to create ultra-broadband and highly efficient fiber-to-chip couplers with low polarization dependence [33]. SWG-based MMI [34] and directional couplers [35] can also be made more compact and can operate over much wider bandwidths than their conventional counterparts. Furthermore, the engineered anisotropy of SWGs is ideal for creating compact and efficient polarization splitters and rotators [36]. SWGs also allow for precise control over the waveguide's group velocity dispersion, a critical parameter for nonlinear optics and for compensating dispersion in communication links.

This approach marks a conceptual leap from geometric optics to metamaterial engineering on a chip. Instead of being constrained by the intrinsic properties of bulk silicon, designers can now synthesize a material with the exact optical properties required for a specific application. While introducing nanoscale features, these waveguides can maintain low propagation losses comparable to standard strip waveguides, with values around 3.7 dB/cm reported for metamaterial-clad structures designed to reduce crosstalk [37].

3. Light Coupling and Splitting: The Interface to and Within the Chip

Efficiently coupling light from an external source, typically a single-mode optical fiber, onto the photonic chip and subsequently distributing that light to various functional blocks are fundamental requirements for any PIC. The performance of these coupling and splitting components directly impacts the overall insertion loss and power budget of the entire system. This section reviews the state of the art in fiber-to-chip couplers and on-chip power splitters.

3.1. Fiber-to-Chip Couplers

A significant challenge in silicon photonics is bridging the immense size mismatch between the mode field diameter of a standard single-mode optical fiber (~9–10 μm) and that of a sub-micron silicon waveguide (~0.5 μm). Two primary strategies have been perfected to address this challenge: grating couplers and edge couplers.

3.1.1. Grating Couplers (GCs)

Grating couplers are periodic structures, such as etched lines in the silicon waveguide, that act as a diffraction grating. They are designed to scatter light out of the plane of the chip at a specific angle that matches the acceptance angle of an optical fiber positioned vertically above it [38]. A key advantage of this out-of-plane coupling scheme is that it allows for testing at any point on a wafer's surface without requiring the wafer to be diced. This makes GCs ideal for high-throughput, wafer-scale characterization of devices during R&D and manufacturing [39].

The performance of GCs is defined by their coupling efficiency (CE) and optical bandwidth. Historically, the CE of simple GCs was limited to around 30–40% (−5 to −3 dB) due to two primary loss mechanisms: back-reflection into the waveguide and power leakage downwards into the silicon substrate. Modern GC design has largely overcome these limitations through sophisticated engineering. Techniques include adding a bottom reflector (e.g., a metal layer or a distributed Bragg reflector) to redirect substrate-bound light upwards, and apodizing the grating (varying the etch depth or duty cycle along its length) to shape the diffracted beam into a more Gaussian-like profile that better matches the fiber mode. More recently, inverse design methods have been used to create non-periodic "meta-gratings" that achieve unprecedented performance. State-of-the-art designs have demonstrated experimental CEs of 78% (1.08 dB) [40] and 81.57% (1.76 dB) [41]. For polarization-diverse applications, two-dimensional (2D) GCs have been developed, with a recent foundry-compatible design achieving a measured CE of 2.54 dB (56%) [42].

Standard uniform GCs are inherently wavelength-sensitive, typically having a 1 dB bandwidth of 30–40 nm. However, advanced designs that engineer the grating's dispersion can significantly broaden this operational window. Bandwidths of 80–90 nm (1 dB) [39] and 92 nm (3 dB) [41] are now achievable. While most GCs are on SOI, SiN platforms are also used. The lower index contrast of SiN results in weaker grating strength and typically lower CE. However, hybrid platforms that use an amorphous silicon or polysilicon overlay on top of the SiN waveguide can enhance the grating strength, enabling CEs approaching 1 dB [43].

3.1.2. Edge Couplers (ECs)

Edge couplers facilitate in-plane coupling, where the optical fiber is brought to the polished facet of the chip. The core component of an EC is a mode-size converter, most commonly an inverted taper. This is a section of the waveguide that gradually narrows towards the chip facet. As the waveguide width decreases, the optical mode becomes less confined and expands in size, allowing for a much better mode-field overlap with the larger mode of the optical fiber [44].

Edge couplers are renowned for their superior performance in terms of efficiency and bandwidth. ECs can achieve near-unity coupling efficiency. By using multiple stacked tapers or multi-tip structures to precisely shape the output mode in both the horizontal and vertical dimensions, losses can be minimized, as shown in Figure 4. Recent demonstrations on the SiN platform have achieved simulated CEs of over 90% across a 1000 nm wavelength range (1000–2000 nm) and a record-high measured CE of 97.1% (−0.13 dB) [45]. On other platforms like thin-film lithium niobate (TFLN), CEs of −1.52 dB (70%) are reported [46]. Furthermore, a major advantage of ECs is their inherently ultra-broad bandwidth and low polarization dependence compared to GCs [45]. A 1 dB bandwidth of over 270 nm has been demonstrated [47], and polarization-dependent loss (PDL)—the difference in coupling loss between TE and TM polarizations—can be as low as < 0.17 dB [48]. Moreover, power handling that can be as high as 520 mW with <0.6/0.8 dB TE/TM insertion loss has been demonstrated recently [49].

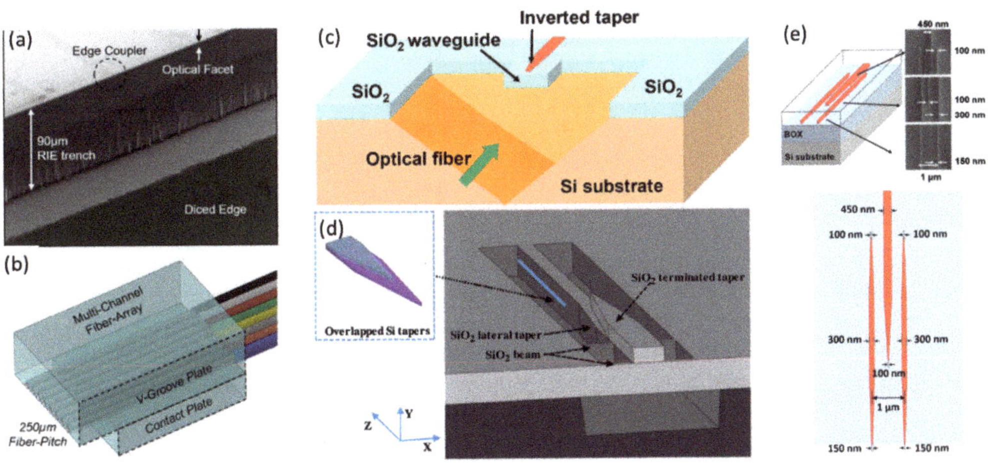

Figure 4. (**a**) SEM image of the optical facet and edge-coupler region on an SOI-PIC; (**b**) schematic of a multichannel fiber array; (c) schematic of the SOI edge-coupling structure based on the use of a Si inverse taper and a waveguide implemented in the BOX; (**d**) schematic of the SOI edge-coupling structure based on the use of a double-layer Si inverse taper and a waveguide; (**e**) schematic of the SOI trident EC structure. Reprinted with permission from [50].

The historical trade-off between the wafer-scale testing convenience of GCs and the superior performance of ECs is rapidly diminishing. As both architectures converge towards high efficiencies, the choice of coupler is becoming less about a performance compromise and more about a system-level design choice dictated by packaging and testing strategy, as shown in Table 2. This convergence is a hallmark of a maturing technology platform.

Table 2. Optical I/O: the edge coupler and grating coupler performance.

Ref	Type	Coupling Eff. per Facet (dB)	Wavelength	Platform	Mode
[46]	Bident EC	1.52 @1.55 μm	Cband	TFLN	TE/TM
[42]	2D GC	2.54	1510–1560 nm	SOI	TE
[45]	Double-tip EC	0.12/0.9 for TE/TM	1070–1830 nm	SIN	TE/TM
[47]	3D multi-tip EC	0.0682/0.0958 for TE/TM @1.55 μm	Cband	TFLN	TE/TM
[38]	GC	0.97/1.05 for TE/TM	Cband	TFLN	TE/TM
[48]	3D EC	0.29/0.24 for TE/TM	1490–1670 nm	TFLN	TE/TM
[41]	Vertical metal GC	2.77/2.55 for TE/TM	Cband	SOI	TE/TM
[44]	Double-tip EC	1.52/2 for TE/TM	Oband	SOI	TE/TM
[49]	EC	0.6/0.8 for TE/TM	Oband	SIN	TE/TM

3.2. On-Chip Power Splitters and Routers

Once light is on the chip, it must be divided and routed to different functional units. The primary components for this task are Y-branches, multimode interference couplers, and directional couplers.

3.2.1. Y-Branch Splitters

The Y-branch is the simplest and one of the most robust power splitters, consisting of a single input waveguide that symmetrically bifurcates into two output waveguides [51]. While conventional Y-branches with a sharp, wide-angle split suffer from significant optical radiation loss, modern designs have largely mitigated this issue. By using a very gradual, adiabatic split or by inserting a specially designed mode-evolution region at the junction, excess loss (the loss beyond the ideal 3 dB split) can be made negligible.

An optimized silica-on-silicon Y-branch using a tapered multimode section reported an excess loss of just 0.12 dB [52]. A design on SiN, optimized for low-loss quantum applications, achieved a loss of <0.1 dB [53]. Additionally, a power splitter based on adiabatically tapered rib waveguides, which functions as a mode evolution Y-branch, demonstrated an exceptionally low excess loss of 0.06 dB [54].

3.2.2. Multimode Interference (MMI) Couplers

MMI couplers are based on the self-imaging principle. Light from one or more input waveguides enters a much wider multimode waveguide section. Within this section, the various excited modes propagate at different phase velocities, interfering with each other. At specific lengths along the MMI section, the interference pattern reproduces one or more images of the input field at the output waveguides [55]. MMI devices are prized for their compact size, broad optical bandwidth, and high tolerance to fabrication variations. As shown in Figure 5, a Si-based 4 × 4 MMI was demonstrated with 100G channel width, high channel uniformity with less than 1 dB excess loss, and 12.7 dB crosstalk in [56].

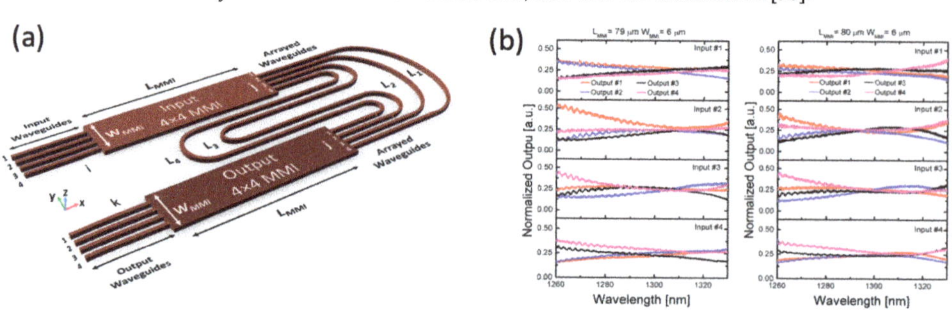

Figure 5. (**a**) Schematic of a 4 × 4 MMI. (**b**) Wavelength-dependent output transmission spectrum from various input ports.

Recent low-loss 2 × 2 MMI couplers fabricated on a SiN platform achieved a measured insertion loss of only 0.07 dB at 1310 nm, with loss remaining below 0.25 dB across the entire O-band [57]. A 1 × 4 MMI splitter on SiN reported a similarly low insertion loss of 0.13 dB [55].

3.2.3. Directional Couplers (DCs)

A directional coupler consists of two parallel waveguides placed in close proximity. Their evanescent fields overlap, allowing power to be gradually transferred from one waveguide to the other. The amount of power transferred (the coupling ratio) depends sensitively on the interaction length, the gap between the waveguides, and the wavelength [54].

While this wavelength sensitivity can be a drawback, it can also be exploited for wavelength-division multiplexing. Moreover, recent advanced designs have focused on overcoming this limitation to create broadband splitters with tailorable coupling ratios. By using bent directional couplers, where the waveguides are curved instead of straight in the coupling region, the wavelength dependence can be significantly flattened. A bent DC on a SiN platform demonstrated a 1 dB bandwidth of 80 nm [20]. Similarly, a broadband directional coupler with an optimized arc-shaped waveguide demonstrated a low excess loss of 0.272 dB. This design maintained a tapping ratio deviation of only 1.433% from the 1550 nm value across the 1470 nm to 1630 nm wavelength range [54], as shown in Figure 6. Furthermore, by using SWG waveguides to form the coupler, the dispersion properties can be engineered to achieve an extremely flat spectral response. A recent SWG-based DC showed an average excess loss of less than 0.24 dB and a coupling ratio variation of less than ±1 dB over an unprecedented 170 nm bandwidth, completely covering the C, L, and U telecommunication bands [58].

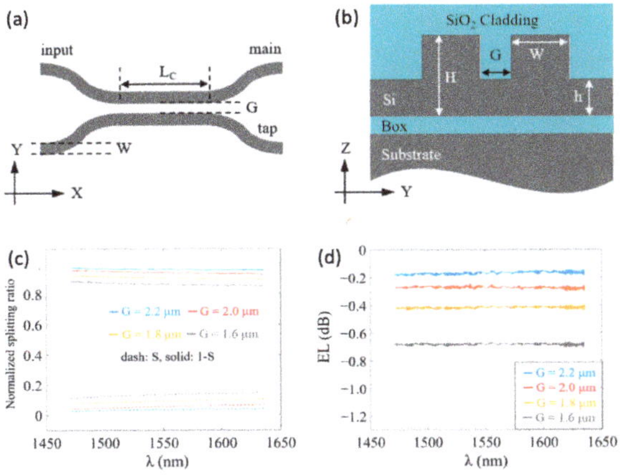

Figure 6. (**a**) Top view of a conventional DC. (**b**) Cross-sectional schematic of rib waveguides in the coupling region. (**c**) Normalized splitting ratio of DCs with different gaps. (**d**) Measured excess loss of DCs with different gaps. Reprinted with permission from [55].

This evolution in splitter design, from simple geometries to more physically sophisticated structures using mode evolution and metamaterial engineering, reflects a deeper understanding of the underlying wave optics. Designers are no longer simply accepting the limitations of basic shapes but are actively manipulating light propagation to achieve the robust, broadband, and fabrication-tolerant performance required for complex, next-generation PICs, as shown in Table 3.

Table 3. Couplers: Y-brand, MMI, and DC performance.

Ref	Type	Coupling Efficiency per Facet (dB)	Wavelength	Platform	Mode
[20]	DC	0.25	1510–1590 nm	SIN	TE
[55]	1 × 4 MMI	0.13	Oband	SIN	TE
[58]	DC	0.24	C + L + Uband	SOI	TE
[57]	2 × 2 MMI	0.07 @1.31 μm	Oband	SIN	TE
[54]	DC	0.272 @1.55 μm	S + C + Lband	SOI	TE
[36]	DC	0.42/0.28 for TE/TM	C + Lband	SOI	TE/TM
[53]	Y-BRANCH	0.1 @808 nm	760–860 nm	SIN	TE
[52]	Y-BRANCH	0.06 @1.55 μm	1400–1565 nm	SOI	TE/TM
[34]	2 × 2 MMI	0.08 @1.55 μm	Cband	SOI	TE
[35]	DC	0.1 @2 μm	1.95–2.02 μm	SOI	TE

4. Wavelength Filtering and Routing: Manipulating the Spectrum

Beyond guiding and splitting light, a critical function of passive photonics is to manipulate the spectral content of the optical signal. Wavelength-selective components are the building blocks for wavelength-division multiplexing (WDM) systems, which dramatically increase the data capacity of a single optical fiber by transmitting multiple data channels on different colors of light. These components are also fundamental to on-chip spectroscopy for sensing and are used to build tunable lasers and other advanced devices.

4.1. Resonant Structures: Microring Resonators (MRRs)

Microring resonators are one of the most versatile and widely used components in silicon photonics. An MRR consists of a closed-loop waveguide, often in the shape of a ring or a "racetrack" with straight sections, that is placed in close proximity to one or two straight "bus" waveguides [59–61]. Light from an input bus waveguide couples evanescently into the ring. It then circulates within the resonator, and resonance occurs when the round-trip path length of the ring is an integer multiple of the wavelength of the light. At these specific resonance wavelengths, light builds up within the cavity, leading to a sharp drop in transmission at the through-port and a corresponding peak at the drop-port (in an add-drop filter configuration).

The performance of an MRR is defined by two key metrics: Quality Factor (Q-factor) and Free Spectral Range (FSR). The Q-factor is a measure of the sharpness of the resonance peak, defined as the ratio of the resonance wavelength to its full width at half-maximum (FWHM) linewidth ($Q = \lambda res/\Delta\lambda FWHM$). A high Q-factor signifies low loss within the cavity and a long photon lifetime, which is essential for creating narrow-band filters, highly sensitive sensors, and low-threshold nonlinear devices. The intrinsic Q-factor is fundamentally limited by the propagation loss of the waveguide forming the ring. On standard SOI platforms, fabrication techniques like local oxidation of silicon (LOCOS), which avoids plasma etching of the waveguide, have enabled intrinsic Q-factors of ~2 million [62]. On the ultra-low-loss SiN platform, Q-factors have reached even more extraordinary levels, with demonstrations of 17 million [63] and a record 45.7 million being reported [64]. A significant recent advancement in integrated photonics is an all-waveguide Si_3N_4 resonator with a record-high intrinsic Q-factor of 422 million at 1570 nm, as shown in Figure 7. This performance is achieved by carefully reducing scattering and absorption losses, resulting in a corresponding linear loss of just 0.060 dB/m—the lowest reported to date for this type of device [65]. This ultra-high-Q capability is also being extended to other wavelength bands; for mid-infrared (MIR) applications, SiGe-on-silicon ring resonators have recently achieved Q-factors of 1 million [66].

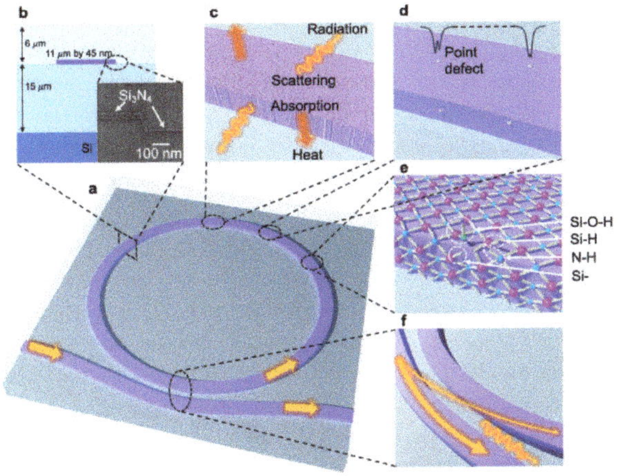

Figure 7. (**a**) Illustration of the resonator design. (**b**) Waveguide design. (**c**) Waveguide surface roughness scatters the guide mode energy into radiation mode and bulk absorption generates heat. (**d**) Point defects split resonances. (**e**) Defect bonds such as Si-O-H, Si-H, N-H, and dangling bonds lead to surface absorption. (**f**) The bus-ring coupler scatters energy into radiation modes and adds excess loss. Reprinted with permission from [65].

The FSR is the wavelength spacing between adjacent resonance peaks. It is inversely proportional to the total optical path length of the ring, given by the equation $FSR = \frac{n_g L}{\lambda^2}$, where L is the physical length of the ring and ng is the group index of the waveguide mode. A large FSR is often desirable to ensure that only one resonance peak falls within the communication band of interest, avoiding ambiguity. The FSR is primarily controlled by the ring's radius. To achieve a large FSR for applications like coarse WDM, very small rings are required. For example, a compact ring on a TFLN platform with an equivalent radius of just 15 μm achieved a large FSR of 10.36 nm [67]. Conversely, for applications requiring very fine channel spacing, large rings are used. As designers seek to balance these competing requirements, a variety of trade-offs emerge between key metrics such as Q-factor, loss, and FSR. These trade-offs, and how different research efforts have addressed them, are summarized in Table 4. A key innovation for improving device yield is the design of FSRs that are insensitive to fabrication variations. This can be achieved by constructing the ring from two sections of waveguide with different widths, engineered such that their group index sensitivities to width variations cancel each other out [68].

Table 4. Filters: mircroring resonator performance.

Ref	Q	Loss (dB/cm)	FWHM	Radius	FSR	Wavelength	Platform	Mode
[65]	422 M	0.0006	906 kHz	11.787 mm	2.713 GHz	1.55–1.65 μm	SIN	TE
[62]	2 M	0.3	1.8 pm	50 μm	2 nm	Cband	SOI	TE
[64]	45.7 M	0.018	0.1 pm (12.5 MHz)	* 100 μm–4 mm	0.52 nm	Cband	SIN	TE/TM
[15]	2 M	0.25	1.76 pm	** 39 μm	0.6 nm	2–2.4 μm	SOI	TE
[63]	17 M	0.021	0.28 pm (35 MHz)	100 μm	25 GHz	1.52–1.6 μm	SIN	TE/TM
[66]	1.01 M	0.15	0.5–1.5 MHz	250 μm	NA	3.5–4.6 μm	SIGE	TE
[67]	48 K	8.5	0.07 nm (8.8 GHz)	15 μm	10.36 nm	1.5–1.6 μm	LNOI	TE

* Euler bends gradual width change. ** Effective Euler bends radius.

4.2. Interferometric Devices: Mach–Zehnder Interferometers (MZIs)

The Mach–Zehnder interferometer is another fundamental building block for spectral filtering. A basic MZI splits an input optical signal into two separate arms, introduces a path length difference between them, and then recombines them [69]. The phase difference accumulated due to the path length difference causes wavelength-dependent constructive and destructive interference at the output, resulting in a sinusoidal transmission spectrum. MZIs are widely used as optical filters and (de)multiplexers. They also form the basis of most thermo-optic switches and high-speed electro-optic modulators, where the phase in one or both arms is actively controlled.

The Extinction Ratio (ER) is a key performance metric for filtering applications, as it measures the ratio of maximum to minimum transmission and is critical for rejecting unwanted signals. While a single MZI has a sinusoidal response, more complex, box-like filter shapes with high ER can be achieved by cascading multiple MZIs or by coupling MRRs to the MZI arms. A recent demonstration of four cascaded MZI-coupled resonators on a SiN platform achieved an experimental extinction ratio of 35 dB [70].

A major challenge for silicon-based interferometric devices is their high sensitivity to temperature, a consequence of silicon's large thermo-optic coefficient ($\sim 1.86 \times 10^{-4}$/K). A temperature change will alter the refractive index and thus the phase in the MZI arms, causing the filter spectrum to drift. A breakthrough solution to this problem is the athermal MZI [71]. In this design, one arm of the interferometer is made from a silicon waveguide, while the other arm is made from a SiN waveguide. Since SiN has a much smaller and positive thermo-optic coefficient, the lengths of the two arms can be chosen to precisely balance out the phase shifts induced by temperature changes. This approach has been shown to reduce the thermal wavelength drift by a factor of 24 compared to an equivalent all-silicon MZI, making the device robust to environmental temperature fluctuations. This shift in focus from achieving peak performance to ensuring robust performance is a sign of the technology's maturation towards commercial deployment.

4.3. Channel Demultiplexing: Arrayed Waveguide Gratings (AWGs)

For applications requiring the separation of many closely spaced wavelength channels, such as in dense WDM (DWDM) systems, the arrayed waveguide grating is the device of choice [72]. An AWG functions like an on-chip prism. It consists of two free-propagation regions (also called star couplers) connected by an array of channel waveguides. Each waveguide in the array has a slightly different length, with a constant length increment between adjacent waveguides.

Light from a single input waveguide enters the first free-propagation region and is distributed among all the arrayed waveguides. As the light propagates through the array, the length difference introduces a wavelength-dependent phase shift. At the second free-propagation region, these phase-shifted beams interfere and focus at different spatial positions depending on their wavelength. A set of output waveguides is placed at the focal plane to collect the separated wavelength channels.

AWGs are capable of high performance, with key metrics being insertion loss, crosstalk, and channel uniformity. State-of-the-art AWGs demonstrate very low loss and excellent channel isolation. A SiN AWG operating in the visible spectrum (760 nm) achieved an insertion loss of less than 0.5 dB and crosstalk below −23 dB [73]. In the 2 μm band, silicon AWGs have been demonstrated with an on-chip loss of 0.5 dB and crosstalk of −30.2 dB [74]. Even very high-channel-count devices, such as a 64-channel AWG with 50 GHz spacing, have been realized with a total on-chip loss of ~5 dB [75].

A persistent challenge in AWG design is the channel non-uniformity, where the outer channels inherently have higher loss than the central channels due to the diffraction envelope of the star coupler. This can be a significant issue in systems where all channels must have similar power levels. A recent innovation to address this is the use of passive silicon nanowires placed in the gaps between the arrayed waveguides at the input to the free-propagation region. These nanowires modify the field distribution, flattening the diffraction envelope and dramatically improving uniformity. This technique was shown to improve the non-uniformity of a 16-channel AWG from 1.6 dB down to 0.63 dB without introducing any additional insertion loss [72]. In addition, Table 5 provides a summary of the key performance metrics for each design.

Table 5. Filters: array waveguide gratings (AWG) performance.

Ref	# of Channels	Loss (dB)	Channel Spacing	Crosstalk	Wavelength	Platform	Mode
[72]	8&16	8 CHs: 6.26, 16 CHs: 10.1	200 GHz	NA	Cband	SOI	TE
[56]	4	0.66	100 GHz	9.6	Oband	SOI	TE
[74]	8	0.5	1.85 nm	30.2	1.88–2.02 μm	SOI	TE
[73]	8	0.5	0.5 nm	23	centered@ 760 μm	SIN	TE
[75]	64	5	0.7 nm	10	1.967–2.012 μm	SOI	TE

5. Foundry Comparison for Fabrication

The remarkable progress in silicon photonic device performance and complexity has been enabled by a corresponding maturation of the manufacturing ecosystem. The high capital cost of a semiconductor fabrication facility makes it prohibitive for most research groups and startup companies to own their own fab. The solution to this has been the adoption of the foundry model, which has democratized access to state-of-the-art fabrication and has been a critical catalyst for innovation.

5.1. The Multi-Project Wafer (MPW) Model

The shift to larger wafer sizes, from 200 mm to 300 mm, is a critical step for manufacturability and cost reduction. Larger wafers allow for a greater number of chips to be fabricated in a single run, which directly lowers the cost per chip, a crucial factor for moving from a research-and-development environment to high-volume manufacturing. While some foundries still operate on 200 mm wafers, the industry trend is moving toward 300 mm, a standard widely adopted by the microelectronics industry. The availability of multiple foundries and the compatibility of their platforms are also key to market growth. While each foundry has its own unique process design kit (PDK), which defines the available components and design rules, standardization efforts are underway to enable greater interoperability and a more robust supply chain. This is vital for ensuring that a design can be easily ported between different fabrication facilities, reducing reliance on a single source.

The Multi-Project Wafer (MPW) service is the cornerstone of the fabless silicon photonics ecosystem [76]. In an MPW run, multiple users—from university research groups to small- and medium-sized enterprises—can purchase a portion of the area on a single wafer. These different projects share the significant costs of the photomask set and the wafer processing, dramatically reducing the financial barrier to entry for prototyping and low-volume production [25,76,77].

This model, as shown in Figure 8, has been instrumental in transforming silicon photonics from a specialized research field into a broadly accessible technology with a clear path to commercialization. By providing regular, scheduled access to a standardized and well-characterized process flow, the MPW model allows designers to fabricate and test their ideas quickly and affordably. Organizations such as ePIXfab in Europe and CMC Microsystems in North America, in partnership with research institutes and foundries, coordinate these shuttle runs, providing a vital service to the community.

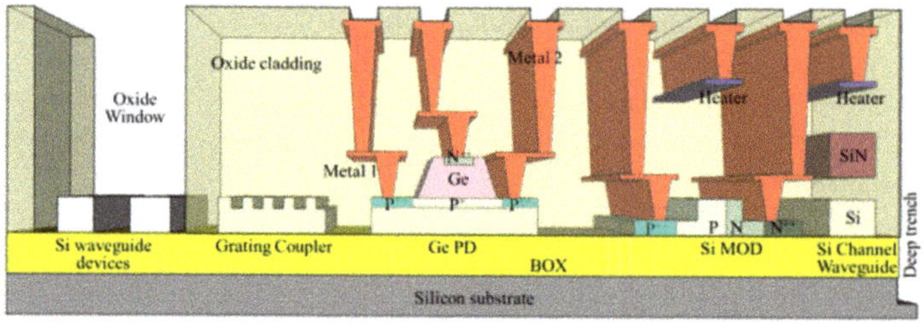

Figure 8. A standard foundry MPW flow. Some common elements are shown, such as phase shifters, photodiodes, and gratings. Reprinted with permission from [76].

5.2. Comparative Overview of Major Foundries

A number of research institutes and commercial semiconductor foundries now offer silicon photonics MPW and dedicated wafer services. Each has a unique platform with distinct capabilities, process flows, and device libraries, catering to different applications and markets. The following provides a comparative overview of several leading foundries.

AIM Photonics: Positioned as a U.S.-based national hub for integrated photonics, AIM Photonics operates an open-access R&D and manufacturing facility focused on 300 mm wafers. Its key strength lies in providing a comprehensive and flexible MPW platform that accelerates the lab-to-fab transition [25]. The platform includes five distinct process technologies: Active, Passive, Low-Loss Active, Sensor, and the specialized Quantum Flex (QFlex) for quantum applications. A unique feature is the offering of "bite-size custom" process modifications, allowing designers to make minor changes to the standard flow for an additional fee, a level of flexibility not typically available at high-volume commercial foundries. The platform integrates silicon, silicon nitride (including thick films >700 nm via a Damascene process for nonlinear optics), and germanium for detectors, targeting applications from datacom and LiDAR to sensing and quantum computing [27].

IMEC: A world-leading research and innovation hub in nanoelectronics based in Belgium, imec offers one of the most advanced and comprehensive silicon photonics platforms, known as iSiPP. Services are available on both 200 mm and 300 mm wafers and include prototyping via MPW runs and volume production partnerships [78–80]. The platform is offered in several flavors, including the simplified iSiPP50G for transceivers operating at 50 Gb/s and beyond, and the full-featured iSiPP200.75. Key process features include a 220 nm SOI starting substrate, multi-level silicon and polysilicon patterning, germanium epitaxy for high-speed photodetectors, and two levels of copper metallization. A major recent upgrade is the co-integration of high-quality, low-loss SiN waveguides without degrading the performance of active devices. Furthermore, imec offers advanced heterogeneous integration capabilities, including the flip-chip bonding of III-V lasers and micro-transfer printing of materials like lithium niobate (LNO) [81].

TSMC (Taiwan Semiconductor Manufacturing Company): As the world's largest semiconductor foundry, TSMC's entry into silicon photonics signifies the technology's industrial maturity. TSMC offers a high-volume SiPh platform on 300 mm wafers, built upon its mature 65 nm CMOS process technology [82–84]. The company's focus is on large-scale production and advanced packaging solutions to meet the demands of the HPC and AI markets. A key offering is the Compact Universal Photonic Engine (COUPE™), a technology that enables the 3D stacking of photonic ICs (PICs) with electronic ICs (EICs), which is essential for co-packaged optics. Their platform includes low-loss SiN waveguides fabricated with a low-temperature (<400 °C) PECVD process, making them compatible with back-end-of-line (BEOL) integration with the CMOS electronics.

GlobalFoundries (GF): GF provides a highly differentiated platform, GF Fotonix™, which monolithically integrates silicon photonics with its RF-CMOS process technology (e.g., 45SPCLO based on 45 nm SOI) [7,85,86]. This allows for the fabrication of both photonic devices and high-performance radio frequency electronics on the same chip. The platform is highly flexible, offering a photonics-only flow as an alternative to the full RF-photonic integration. The second generation of GF Fotonix supports data rates of 200 G per wavelength and includes upgrades to all active devices. GF is also investing heavily in full turnkey solutions, including onshore advanced packaging and testing, to provide customers with a secure and complete supply chain from chip fabrication to the final packaged product. The platform includes SiN for power handling and through-silicon vias (TSVs) for 3D stacking.

Tower Semiconductor: Tower offers its PH18 silicon photonics platform, which is promoted as an "open" foundry service available to all customers for prototyping and volume production on 200 mm wafers [87,88]. This platform is designed to complement Tower's strong offerings in specialty analog processes like SiGe BiCMOS, providing a combined electronic–photonic solution for customers. The PH18 platform features low-loss silicon and silicon nitride waveguides, integrated germanium photodiodes (PIN and avalanche), and efficient heaters for thermal tuning. A key strength is its comprehensive design enablement, with process design kits (PDKs) supported by all major electronic design automation (EDA) vendors, including Ansys, Cadence, and Synopsys. The platform primarily targets the O-band and C-band data center interconnect market, as well as telecom and sensing applications. The summary table is shown in Table 6.

Table 6. Summary of representative SiPh foundry platforms.

Foundry	Platform Highlights	Key Materials	Process Node(s)/Wafer Size
AIM Photonics	US-based open-access R&D hub; Comprehensive MPW platform (Active, Passive, Low-Loss, Sensor, QFlex); "Bite-size" process customization.	Si, SiN (incl. thick film), Ge	90 nm/45 nm base on 300 mm
IMEC	iSiPP platforms (iSiPP50G, iSiPP200); Advanced co-integration of SiN; Heterogeneous III-V/LNO integration via flip-chip/micro-transfer printing.	Si, poly-Si, SiN, Ge, III-V (InP), LNO	130 nm base on 200 mm, 300 mm
TSMC	High-volume 65 nm SiPh process; COUPE™ advanced packaging for 3D stacking; Fully automated wafer-level testing (EWAT/OWAT).	Si, SiN (low-temp PECVD)	65 nm on 300 mm
GlobalFoundries	GF Fotonix™ platform; Monolithic RF-CMOS integration; Turnkey solutions including advanced packaging; 200 G/λ capable.	Si, SiN, Ge	90 nm, 45 nm on 200 mm/300 mm
Tower Semiconductor	PH18 "open" foundry platform on 200 mm; Complements SiGe BiCMOS processes; Strong PDK support from major EDA vendors.	Si, SiN, Ge	180 nm base on 200 mm

This table provides a general overview of silicon photonics foundry capabilities and should be used for reference only. Please note that the specifications and offerings are subject to change. For the most up-to-date and representative information, it is essential to consult the official process design kits (PDKs) and documentation from each foundry, as their state-of-the-art technologies are constantly evolving.

5.3. Integration with Active Devices: Approaches for System-Level Functionality

While this review focuses on the foundational passive components that form the backbone of a photonic integrated circuit, a complete system requires the integration of active devices that can modulate, detect, and generate light. The integration of these active components—such as modulators, photodetectors, lasers, and amplifiers—onto the passive silicon platform is a major challenge and a critical area of ongoing research. Three primary approaches are used to achieve this system-level functionality: monolithic, heterogeneous, and hybrid integration.

5.3.1. Monolithic Integration

Monolithic integration involves fabricating both the electronic and photonic devices on a single silicon substrate using a single, unified process flow. This approach is highly compatible with the mature CMOS manufacturing ecosystem [85]. It is particularly well-suited for high-speed modulators [89] and photodetectors [90], which can be formed using doped silicon and germanium. For example, silicon's Pockels effect is weak, but its high plasma dispersion effect allows for the creation of efficient modulators. Germanium, with its high absorption at telecom wavelengths, can be epitaxially grown on silicon to create high-speed photodetectors. This method offers the potential for very low power consumption and high-density integration because the electrical interconnects between the electronic and photonic components are kept extremely short. This approach also simplifies packaging, as there are no separate chips to align. A key challenge remains in the lack of a native silicon-based laser source, as silicon is an inefficient light emitter due to its indirect bandgap. However, efficient laser-to-SiPh-circuit butt-coupling with optical power up to 20 mW was demonstrated through wafer- and module-level characterizations [91].

5.3.2. Heterogeneous Integration

To overcome the light source limitation, heterogeneous integration allows for the bonding of III-V semiconductor materials (such as InP and GaAs, which are efficient light emitters) directly onto SOI wafer [92]. The III-V material is typically grown on its native substrate and then transferred and bonded to the passive silicon waveguide layer using a process like wafer bonding. The transferred material is then patterned to create active devices like lasers, semiconductor optical amplifiers (SOAs), and modulators. This technique leverages the best of both worlds: the high-performance light-emitting capabilities of III-V materials and the low-loss, cost-effective passive routing of the silicon platform [93]. This is the most common approach for creating integrated laser sources on silicon. A significant advantage is that it allows for the parallel processing of both the silicon and III-V wafers, which can improve manufacturing yield.

5.3.3. Hybrid Integration

Hybrid integration involves placing separate, pre-fabricated III-V and silicon photonic chips side-by-side or on a common substrate and then connecting them [94,95]. This can be performed via techniques such as flip-chip bonding. While this method can lead to higher insertion loss and a larger footprint than heterogeneous or monolithic integration, it offers the highest level of flexibility and yield. It allows designers to select best-in-class components from different processes (e.g., a high-power III-V laser from one vendor and a low-loss silicon PIC from another) and then assemble them at the package level. This approach is particularly relevant for advanced systems like co-packaged optics (CPO), which integrate both the electronic IC (EIC) and the photonic IC (PIC) onto a single package, minimizing the length of high-speed electrical traces.

Therefore, the choice of integration approach is a key consideration in system design, trading off between manufacturability, performance, and cost. While monolithic integration offers the highest level of integration density, heterogeneous and hybrid methods are vital for incorporating high-performance III-V devices, such as laser sources, that are essential for a complete and functional silicon photonic system.

6. Challenges and Future Outlook

Despite the tremendous progress in the performance and manufacturing of passive silicon photonic devices, several significant challenges remain. Overcoming these hurdles is essential for the technology to realize its full potential. Concurrently, a number of exciting trends are emerging that promise to redefine the capabilities and application scope of silicon photonics, pushing the field into new and uncharted territories.

6.1. Persistent Technical Hurdles

6.1.1. Polarization Management

One of the most persistent challenges in silicon photonics stems from the material and geometric properties of the waveguides themselves. The high refractive index contrast of the SOI platform and the typically rectangular cross-section of the waveguides lead to a very large birefringence [96]. This means that the effective refractive index experienced by the light is significantly different for the TE mode and the TM mode. As a result, the performance of most photonic devices—from couplers to filters—is highly polarization-dependent. This poses a major problem because the polarization state of light arriving from a standard optical fiber is generally arbitrary and can fluctuate over time.

To address this, on-chip polarization management is required [97–99]. This is typically achieved through a polarization diversity approach, where the incoming light is split into its TE and TM components by a polarization beam splitter (PBS). The TM polarization is then converted to TE polarization using a polarization rotator (PR), and both TE signals are processed in parallel on identical circuits. Recent advancements in these crucial components include the creation of ultra-compact polarization beam rotator-splitters (PBRS) have been created using inverse design, combining both functions into a single, high-performance device [100]. Sub-wavelength gratings can also be engineered to be highly anisotropic, allowing for the design of efficient TM-pass polarizers and broadband PBSs [96,101]. Additionally, tunable polarization management is being explored using micro-electromechanical systems (MEMS). By mechanically perturbing a waveguide or changing the gap in a directional coupler, MEMS structures can enable tunable polarization splitting and rotation, offering reconfigurability that is not possible with static devices [102].

6.1.2. Packaging and Integration

While device fabrication has benefited enormously from leveraging CMOS infrastructure, photonic packaging remains a major bottleneck and a significant contributor to the final cost of a photonic-enabled system [5,8,103]. The challenges are multifaceted, encompassing optical I/O, electrical integration, heterogeneous integration, and thermal management [104].

Achieving robust, low-loss, and high-throughput alignment of optical fibers (or fiber arrays) to the chip's couplers is a critical and often time-consuming step. While passive alignment schemes are being developed, active alignment, where the device is powered on and the fiber position is optimized for maximum coupling, is still common but expensive and slow. High-speed electrical signals must be routed to and from the PIC to drive active components like modulators. This requires careful RF design and integration to avoid signal degradation and crosstalk.

Furthermore, most silicon photonic systems require an external laser source because silicon is an inefficient light emitter. Heterogeneous integration, whether through flip-chip bonding or wafer-level techniques, is a complex assembly process to integrate these III-V semiconductor lasers with the silicon PIC. Lastly, thermal management is crucial for stable operation, as PICs, especially those with densely integrated electronics and active photonic components, can generate significant heat. Silicon photonic devices are highly sensitive to temperature, so thermal crosstalk between components must be carefully modeled and mitigated.

Solutions are being pursued through the development of standardized packaging design rules (PDRs) and the holistic, system-level approach of co-packaged optics.

6.2. Emerging Trends and Future Directions

6.2.1. Co-Packaged Optics (CPO)

Co-packaged optics represents a fundamental shift in how optical interconnects are integrated into high-performance computing systems. The concept involves moving the optical I/O functionality from front-panel pluggable modules directly onto the same package or substrate as the main electronic IC, such as a network switch ASIC or a GPU, as shown in Figure 9. By co-integrating the photonic and electronic chiplets, the electrical distance that high-speed signals must travel is reduced from tens of centimeters across a PCB to mere millimeters or microns on the package substrate [105–107].

This architectural change is driven primarily by the immense power consumption of modern data centers. In a traditional pluggable architecture, a significant portion of the system power (~15 pJ/bit) is consumed by the SerDes (Serializer/Deserializer) circuits needed to drive the long, lossy copper traces to the front panel. CPO promises to slash this interconnect power to <5 pJ/bit, a critical step for enabling the terabit-per-second switches and AI accelerators of the future. However, this approach introduces significant new challenges in thermal management (as the hot ASIC is now next to the temperature-sensitive PIC), manufacturing yield, testing, and field serviceability.

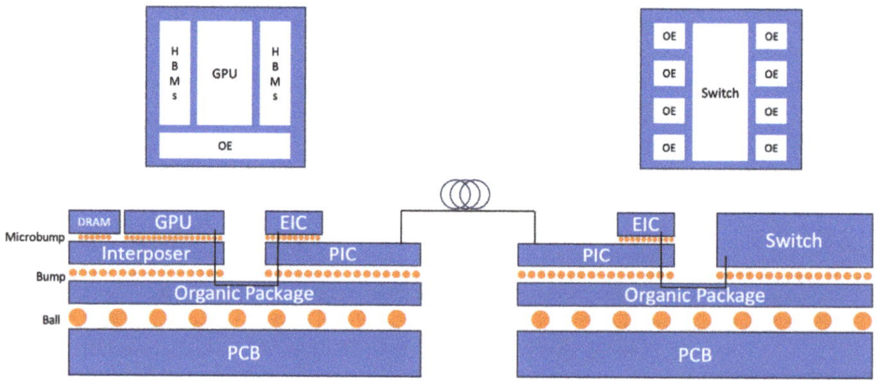

Figure 9. Schematic of silicon photonic CPO technology. OE: optical engine. HBM: high bandwidth memory.

6.2.2. Inverse Design and Artificial Intelligence (AI)

The increasing complexity of photonic devices is driving a revolution in design methodology. The traditional approach, which relies on human intuition and parameter sweeps of known structures, is often too slow and suboptimal for creating the novel components required for next-generation systems. Inverse design is a new paradigm that flips this process on its head [108,109]. Instead of starting with a structure, the designer specifies a desired optical function (a figure of merit), and an optimization algorithm explores a vast design space to generate a device topology that best achieves that function [110].

These algorithms, such as topology optimization (TO) or genetic algorithms (GA), often produce non-intuitive, free-form, or "pixelated" structures that can significantly outperform their conventional counterparts, as shown in Figure 10. This has led to breakthroughs in creating ultra-compact and highly efficient devices like splitters and grating couplers [111].

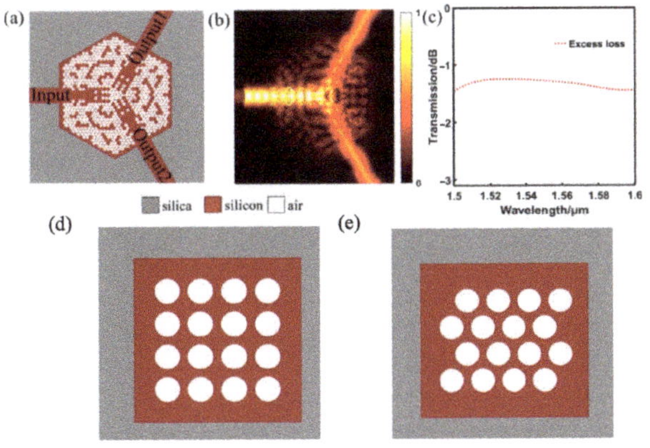

Figure 10. The simulated results of arbitrary-input 1 × 2 power splitter. (**a**) The optimized structure. (**b**) The corresponding distribution of optical field. (**c**) The simulated spectra transmission of this device. (**d**,**e**) Two ways to arrange pixels as inverse design unit.

Complementing inverse design, artificial intelligence and machine learning models are being used to accelerate the design cycle even further [112]. Deep neural networks (DNNs) can be trained on simulation data to act as ultra-fast surrogates for computationally expensive electromagnetic solvers. This allows for the rapid prediction of a device's performance, enabling vast optimizations that would otherwise be intractable. A key challenge that remains is the "black-box" nature of these methods; understanding the underlying physics of the novel structures they produce is an active area of research.

6.2.3. New Frontiers: Quantum and Sensing

The maturation of the silicon photonics platform is enabling its application in entirely new domains beyond communications. Quantum technologies represent a significant frontier, with silicon photonics emerging as a leading platform for building photonic quantum computers. The ability to fabricate large-scale, complex interferometric circuits with high stability and low loss on a chip is ideal for manipulating photonic qubits [113,114]. Photons are robust qubits, resilient to decoherence, and can be operated at room temperature. The scalability and manufacturability offered by CMOS foundries are a major advantage for constructing the future's million-qubit, fault-tolerant systems.

Another key area is advanced sensing, where the high sensitivity of photonic devices to their environment is being leveraged for a new generation of miniaturized, low-cost, and high-performance sensors [115]. Applications include biosensing [116], where high-Q microring resonators are used as highly sensitive label-free biosensors to detect the presence of proteins, viruses, or DNA by measuring minute changes in the refractive index of their surroundings. Spectroscopy is also enabled on-chip through miniaturized Fourier-transform spectrometers (FTS) or AWGs, used for chemical analysis and gas sensing, particularly in the mid-infrared range [117]. Furthermore, LiDAR systems for autonomous vehicles and robotics are being developed using PICs to build solid-state systems that steer laser beams without any moving parts [118].

These emerging trends are not developing in isolation but are part of a virtuous cycle. The demand for more computational power for AI is a primary driver for CPO. The complexity of CPO systems necessitates the use of AI-driven inverse design tools. The resulting ability to design and fabricate highly complex, integrated systems then opens the door to revolutionary applications like large-scale quantum processors and multi-analyte sensor arrays, which in turn create new demands, further accelerating the entire field.

7. Conclusions

7.1. Synthesis of Key Advancements

The field of passive silicon photonics has undergone a period of transformative growth, evolving from a promising academic research area into a mature, industrialized technology. This review has charted the remarkable progress across its foundational components. We have witnessed the propagation loss of optical waveguides plummet from the dB/cm to the dB/m regime, a thousand-fold improvement driven by advances in rib waveguide architectures, ultra-pure silicon nitride films, and novel fabrication techniques like the Damascene process. This ultra-low loss has been the direct enabler for resonant devices, such as microring resonators, to achieve extraordinary quality factors in the tens of millions, unlocking new possibilities in filtering, sensing, and nonlinear optics.

Simultaneously, the critical interfaces to and within the chip have been perfected. Both grating couplers and edge couplers have converged on near-unity coupling efficiencies, effectively erasing a long-standing performance trade-off and giving system architects greater flexibility in packaging and testing strategies. On-chip splitters and routers, once limited by the simple geometries of Y-branches and directional couplers, are now sophisticated components engineered for broadband, low-loss, and fabrication-tolerant operation through the use of mode evolution principles and sub-wavelength grating metamaterials.

Underpinning all these device-level advancements is the maturation of the global fabrication ecosystem. The multi-project wafer (MPW) model has successfully democratized access to state-of-the-art manufacturing, fostering a vibrant community of innovation among universities and fabless design houses. Major commercial and research foundries now offer robust, well-characterized process design kits (PDKs) that include not only silicon but also co-integrated materials like silicon nitride and germanium, alongside advanced packaging solutions, providing a clear and reliable path from concept to high-volume production.

7.2. Concluding Perspective on the Field's Trajectory

Looking at the trajectory of the field, it is clear that silicon photonics is successfully navigating the transition from a focus on optimizing individual component performance to a focus on system-level integration, manufacturability, and robustness. The central questions are evolving from "Can we build a high-performance device?" to "Can we integrate, package, test, and scale these complex systems reliably and cost-effectively?" This shift is indicative of a technology reaching industrial maturity.

The convergence of several powerful trends—the insatiable demand for bandwidth from AI, the architectural shift towards co-packaged optics, the paradigm change in design methodology brought by AI-driven tools, and the expansion into new application frontiers like quantum computing and advanced sensing—creates a powerful, self-reinforcing cycle of innovation. As these elements co-evolve, they accelerate the capabilities of the platform at an exponential rate. Silicon photonics is therefore not merely an incremental improvement over copper interconnects; it is a foundational enabling technology poised to follow a similar trajectory of impact and integration as the electronic integrated circuit industry. It is cementing its role as the essential hardware backbone for the data-intensive, intelligent, and quantum-aware world of the 21st century.

Author Contributions: Resources and Original Draft, and Review, Q.L. and Y.B.; Writing—Review and Editing, J.X.; Project Administration and Supervision, Q.L. All authors contributed to data analysis and manuscript writing. All authors have read and agreed to the published version of the manuscript.

Funding: This research received no external funding.

Institutional Review Board Statement: Not applicable.

Informed Consent Statement: Not applicable.

Data Availability Statement: The data related to the paper are available from the corresponding authors upon reasonable request.

Conflicts of Interest: The authors declare no conflicts of interest.

References

1. Udkar, S. Harnessing Photonic Computing for Next-Generation CPUs and GPUs in High-Performance Computing. *Int. J. Emerg. Trends Comput. Sci. Inf. Technol.* **2024**, *5*, 23–36. [CrossRef]
2. Wang, Z.; Xiao, Y.; Liao, K.; Li, T.; Song, H.; Chen, H.; Uddin, S.M.Z.; Mao, D.; Wang, F.; Zhou, Z.; et al. Metasurface on integrated photonic platform: From mode converters to machine learning. *Nanophotonics* **2022**, *11*, 3531–3546. [CrossRef] [PubMed]
3. Ning, S.; Zhu, H.; Feng, C.; Gu, J.; Jiang, Z.; Ying, Z.; Midkiff, J.; Jain, S.; Hlaing, M.H.; Pan, D.Z.; et al. Photonic-Electronic Integrated Circuits for High-Performance Computing and AI Accelerators. *J. Light. Technol.* **2024**, *42*, 7834–7859. [CrossRef]
4. Resch, M.M.; Gebert, J.; Hoppe, D. The Future of HPC and AI. In Proceedings of the 2025 International Conference on Intelligent Control, Computing and Communications (IC3), Mathura, India, 13–14 February 2025; pp. 897–904. [CrossRef]
5. Margalit, N.; Xiang, C.; Bowers, S.M.; Bjorlin, A.; Blum, R.; Bowers, J.E. Perspective on the future of silicon photonics and electronics. *Appl. Phys. Lett.* **2021**, *118*, 220501. [CrossRef]
6. Rahim, A.; Hermans, A.; Wohlfeil, B.; Petousi, D.; Kuyken, B.; Van Thourhout, D.; Baets, R.G. Taking silicon photonics modulators to a higher performance level: State-of-the-art and a review of new technologies. *Adv. Photonics* **2021**, *3*, 024003. [CrossRef]
7. Hirokawa, T.; Bian, Y.; Giewont, K.; Aboketaf, A.; Chandran, S.; Cho, J.-K.; Chowdhury, Z.; Lee, W.S.; Liu, Q.; Sharma, P.; et al. Latest progress and challenges in 300 mm monolithic silicon photonics manufacturing. In Proceedings of the Optical Fiber Communication Conference, San Diego, CA, USA, 24–28 March 2024; Optica Publishing Group: Washington, DC, USA, 2024; p. Th3H-2. [CrossRef]
8. Shekhar, S.; Bogaerts, W.; Chrostowski, L.; Bowers, J.E.; Hochberg, M.; Soref, R.; Shastri, B.J. Roadmapping the next generation of silicon photonics. *Nat. Commun.* **2024**, *15*, 751. [CrossRef] [PubMed]
9. Dhote, C.; Singh, A.; Kumar, S. Silicon Photonics Sensors for Biophotonic Applications—A Review. *IEEE Sens. J.* **2022**, *22*, 18228–18239. [CrossRef]
10. Debnath, K.; Arimoto, H.; Husain, M.K.; Prasmusinto, A.; Al-Attili, A.; Petra, R.; Chong, H.M.H.; Reed, G.T.; Saito, S. Low-Loss Silicon Waveguides and Grating Couplers Fabricated Using Anisotropic Wet Etching Technique. *Front. Mater.* **2016**, *3*, 10. [CrossRef]
11. Dong, P.; Qian, W.; Liao, S.; Liang, H.; Kung, C.-C.; Feng, N.-N.; Shafiiha, R.; Fong, J.; Feng, D.; Krishnamoorthy, A.V.; et al. Low loss shallow-ridge silicon waveguides. *Opt. Express* **2010**, *18*, 14474–14479. [CrossRef]
12. Vlasov, Y.A.; McNab, S.J. Losses in single-mode silicon-on-insulator strip waveguides and bends. *Opt. Express* **2004**, *12*, 1622–1631. [CrossRef]
13. Gould, M.; Pomerene, A.; Hill, C.; Ocheltree, S.; Zhang, Y.; Baehr-Jones, T.; Hochberg, M. Ultra-thin silicon-on-insulator strip waveguides and mode couplers. *Appl. Phys. Lett.* **2012**, *101*, 221106. [CrossRef]
14. Wu, S.; Mu, X.; Cheng, L.; Mao, S.; Fu, H.Y. State-of-the-Art and Perspectives on Silicon Waveguide Crossings: A Review. *Micromachines* **2020**, *11*, 326. [CrossRef] [PubMed]
15. Wei, J.; Zhu, C.; Yu, Y.; Wang, R.; Yu, S. Low-loss silicon waveguide and an ultrahigh-Q silicon microring resonator in the 2 μm wave band. *Opt. Lett.* **2024**, *49*, 3271–3274. [CrossRef]
16. Tran, M.A.; Huang, D.; Komljenovic, T.; Peters, J.; Malik, A.; Bowers, J.E. Ultra-Low-Loss Silicon Waveguides for Heterogeneously Integrated Silicon/III-V Photonics. *Appl. Sci.* **2018**, *8*, 1139. [CrossRef]
17. Thander, A.K. Numerical methods for modal analysis in optical rib waveguide structures: A review. *J. Opt.* **2024**, 1–7. [CrossRef]

18. Huang, H.; Liu, K.; Qi, B.; Sorger, V.J. Re-Analysis of Single-Mode Conditions for Silicon Rib Waveguides at 1550 nm Wavelength. *J. Light. Technol.* **2016**, *34*, 3811–3817. [CrossRef]
19. Li, Z.; Fan, Z.; Zhou, J.; Cong, Q.; Zeng, X.; Zhang, Y.; Jia, L. Process Development of Low-Loss LPCVD Silicon Nitride Waveguides on 8-Inch Wafer. *Appl. Sci.* **2023**, *13*, 3660. [CrossRef]
20. Li, D.; Li, B.; Tang, B.; Zhang, P.; Yang, Y.; Liu, R.; Xie, L.; Li, Z. Broadband Silicon Nitride Power Splitter Based on Bent Directional Couplers with Low Thermal Sensitivity. *Micromachines* **2022**, *13*, 559. [CrossRef] [PubMed]
21. Chandran, S.; Bian, Y.; Lee, W.S.; Abumazwed, A.; Liu, N.; Xu, L.; Ding, H.; Aboketaf, A.; Rakowski, M.; Dezfulian, K.; et al. High performance silicon nitride passive optical components on monolithic silicon photonics platform. In Proceedings of the Optical Fiber Communication Conference, San Diego, CA, USA, 24–28 March 2024; Optica Publishing Group: Washington, DC, USA, 2024; p. Th3H-4. [CrossRef]
22. Bauters, J.F.; Heck, M.J.R.; John, D.D.; Barton, J.S.; Bruinink, C.M.; Leinse, A.; Heideman, R.G.; Blumenthal, D.J.; Bowers, J.E. Planar waveguides with less than 0.1 dB/m propagation loss fabricated with wafer bonding. *Opt. Express* **2011**, *19*, 24090–24101. [CrossRef]
23. Ji, X.; Wang, R.N.; Liu, Y.; Riemensberger, J.; Qiu, Z.; Kippenberg, T.J. Efficient mass manufacturing of high-density, ultra-low-loss Si_3N_4 photonic integrated circuits. *Optica* **2024**, *11*, 1397–1407. [CrossRef]
24. Ji, X.; Okawachi, Y.; Gil-Molina, A.; Corato-Zanarella, M.; Roberts, S.; Gaeta, A.L.; Lipson, M. Ultra-Low-Loss Silicon Nitride Photonics Based on Deposited Films Compatible with Foundries. *Laser Photonics Rev.* **2023**, *17*, 2200544. [CrossRef]
25. Fahrenkopf, N.M.; McDonough, C.; Leake, G.L.; Su, Z.; Timurdogan, E.; Coolbaugh, D.D. The AIM Photonics MPW: A Highly Accessible Cutting Edge Technology for Rapid Prototyping of Photonic Integrated Circuits. *IEEE J. Sel. Top. Quantum Electron.* **2019**, *25*, 8201406. [CrossRef]
26. Carpenter, L.G.; van Niekerk, M.; Begović, A.; Sundaram, V.S.; Deenadayalan, V.; Palone, T.; Fanto, M.; Preble, S.; Baiocco, C.; Leake, G.L.; et al. Towards Low Propagation Losses in Active Photonic Multi-Project Wafer Runs. In Proceedings of the OSA Advanced Photonics Congress 2021, Washington, DC, USA, 26–29 July 2021; Optica Publishing Group: Washington, DC, USA, 2021; p. ITu3A.5. [CrossRef]
27. Kim, E.; Carpenter, L.; Smith, A.M.; Tison, C.C.; Coleman, D.; Leake, G.; Fanto, M.; Rizzo, A.J.; García-Blanco, S.M.; Cheben, P. Ultra-low loss dispersion-engineered silicon nitride waveguides on 300 mm wafers. In Proceedings of the Integrated Optics: Devices, Materials, and Technologies XXIX, San Francisco, CA, USA, 27–30 January 2025; SPIE: Bellingham, DC, USA, 2025; pp. 271–277. [CrossRef]
28. Pfeiffer, M.H.P.; Herkommer, C.; Liu, J.; Morais, T.; Zervas, M.; Geiselmann, M.; Kippenberg, T.J. Photonic Damascene Process for Low-Loss, High-Confinement Silicon Nitride Waveguides. *IEEE J. Sel. Top. Quantum Electron.* **2018**, *24*, 6101411. [CrossRef]
29. Liu, J.; Huang, G.; Wang, R.N.; He, J.; Raja, A.S.; Liu, T.; Engelsen, N.J.; Kippenberg, T.J. High-yield, wafer-scale fabrication of ultralow-loss, dispersion-engineered silicon nitride photonic circuits. *Nat. Commun.* **2021**, *12*, 2236. [CrossRef]
30. Luque-González, J.M.; Sánchez-Postigo, A.; Hadij-ElHouati, A.; Ortega-Moñux, A.; Wangüemert-Pérez, J.G.; Schmid, J.H.; Cheben, P.; Molina-Fernández, Í.; Halir, R. A review of silicon subwavelength gratings: Building break-through devices with anisotropic metamaterials. *Nanophotonics* **2021**, *10*, 2765–2797. [CrossRef]
31. Bock, P.J.; Cheben, P.; Schmid, J.H.; Lapointe, J.; Delâge, A.; Xu, D.-X.; Janz, S.; Densmore, A.; Hall, T.J. Subwavelength grating crossings for silicon wire waveguides. *Opt. Express* **2010**, *18*, 16146–16155. [CrossRef]
32. Molina-Fernández, I.; Wangüemert-Perez, J.G.; Ortega-Moñuz, A.; Halir, R.; Rubio, J.d.O.; Sánchez-Postigo, A.; Luque-González, J.M.; Hadij-Elhouati, A.; Pereira-Martín, D.; Gonzalez-Andrade, D.; et al. Silicon subwavelength waveguiding devices. In Proceedings of the High Contrast Metastructures IX, San Francisco, CA, USA, 26 February 2020; SPIE: Bellingham, DC, USA, 2020; pp. 114–120. [CrossRef]
33. Kazanskiy, N.L.; Butt, M.A.; Khonina, S.N. Silicon photonic devices realized on refractive index engineered subwavelength grating waveguides—A review. *Opt. Laser Technol.* **2021**, *138*, 106863. [CrossRef]
34. Shi, Y.; Shao, B.; Zhang, Z.; Zhou, T.; Luo, F.; Xu, Y. Ultra-Broadband and Low-Loss Silicon-Based Power Splitter Based on Subwavelength Grating-Assisted Multimode Interference Structure. *Photonics* **2022**, *9*, 435. [CrossRef]
35. Zhu, D.; Ye, H.; Liu, Y.; Li, J.; Wang, Y.; Yu, Z. Silicon subwavelength grating-assisted asymmetric directional coupler around 2 μm and its applications. *Opt. Laser Technol.* **2021**, *136*, 106789. [CrossRef]
36. Chen, J.; Xiao, J. Ultracompact silicon-based polarization splitter and rotator based on asymmetric directional couplers with subwavelength gratings. *JOSA B* **2022**, *39*, 345–354. [CrossRef]
37. Halir, R.; Bock, P.J.; Cheben, P.; Ortega-Moñux, A.; Alonso-Ramos, C.; Schmid, J.H.; Lapointe, J.; Xu, D.; Wangüemert-Pérez, J.G.; Molina-Fernández, Í.; et al. Waveguide sub-wavelength structures: A review of principles and applications. *Laser Photonics Rev.* **2015**, *9*, 25–49. [CrossRef]
38. Sultan, A.; Khalil, M.; Mehravar, L.; Xu, C. Highly Efficient Polarization-Insensitive Grating Couplers on Thin-Film Lithium Niobate with an Integrated Gold Layer. *Photonics* **2025**, *12*, 111. [CrossRef]

39. Cheng, L.; Mao, S.; Li, Z.; Han, Y.; Fu, H.Y. Grating Couplers on Silicon Photonics: Design Principles, Emerging Trends and Practical Issues. *Micromachines* **2020**, *11*, 666. [CrossRef]
40. Tahara, N.; Nawa, S.; Taira, R.; Suyama, S.; Hirotani, K.; Kamata, M.; Maegami, Y.; Tsuchizawa, T.; Yamamoto, N.; Yamada, K.; et al. High efficiency silicon photonics grating coupler compatible to standard foundry service. In Proceedings of the Silicon Photonics XX, San Francisco, CA, USA, 28–30 January 2025; SPIE: Bellingham, DC, USA, 2025; pp. 57–65. [CrossRef]
41. Cheng, H.; Wu, J.; Wang, Y.; Ran, C.; Li, H.; Wang, Y.; Li, Y.; Zhang, S.; Wang, C.; Yang, J. Ultra-Compact, High-Efficiency Vertical Meta-Grating Couplers for Meta-Photonic Integrated Circuits. *Nanomaterials* **2025**, *15*, 583. [CrossRef]
42. Zhou, W.; Lu, K.; Kang, S.; Wu, X.; Tong, Y. Efficient Polarization-Diversity Grating Coupler With Multipolar Radiation Mode Enhancement. *IEEE Photonics J.* **2025**, *17*, 7100206. [CrossRef]
43. Korček, R.; Fraser, W.; Wilmart, Q.; Quiroz, D.M.; Litvik, J.; Zubor, L.; Krehel, T.; Edmond, S.; Milanizadeh, M.; Schmid, J.H.; et al. High-efficiency off-chip grating couplers for silicon nitride photonics. In Proceedings of the Integrated Optics: Devices, Materials, and Technologies XXVIII, San Francisco, CA, USA, 29 January–1 February 2024; SPIE: Bellingham, DC, USA, 2024; pp. 41–45. [CrossRef]
44. Wan, Y.; Zhang, Y.; Wang, J. O-band low loss and polarization insensitivity bilayer and double-tip edge coupler. *Adv. Photonics Nexus* **2025**, *4*, 026004. [CrossRef]
45. Ai, X.; Xuan, M.W.; Zhang, Y.; Hsu, W.-L.; Veilleux, S.; Dagenais, M. Broadband high coupling efficiency edge coupler with low polarization-dependence on the silicon-nitride platform. *Opt. Express* **2025**, *33*, 16253–16262. [CrossRef] [PubMed]
46. Liu, M.; Zhang, T.; Xu, B.; Zhang, Y.; Du, J.; Sun, J.; Wang, J. High-efficiency bident edge coupler based on thin film lithium niobate. *Opt. Express* **2025**, *33*, 16338–16349. [CrossRef]
47. Zhang, T.; Li, J.; Li, M.; Liu, J. Design and Optimization of a High-Efficiency 3D Multi-Tip Edge Coupler Based Lithium Niobate on Insulator Platform. *Photonics* **2024**, *11*, 134. [CrossRef]
48. Chen, B.; Ruan, Z.; Wang, M.; Gong, S.; Liu, L. High-performance and fabrication-tolerant edge coupler on thin film lithium niobate based on a three-dimensional inverse taper. *APL Photonics* **2024**, *9*, 116111. [CrossRef]
49. Bian, Y.; Hirokawa, T.; Karra, V.; Dasgupta, A.; Lee, W.S.; Aboketaf, A.; Afzal, F.; Sporer, R.; Nummy, K.; Giewont, K.; et al. Monolithically integrated self-aligned SiN edge coupler with <0.6/0.8 dB TE/TM insertion loss, <−39 dB back reflection and >520 mW high-power handling capability. In Proceedings of the Optical Fiber Communication Conference (OFC) 2023, San Diego, CA, USA, 5–9 March 2023; Optica Publishing Group: Washington, DC, USA, 2023; p. M3C.3. [CrossRef]
50. Marchetti, R.; Lacava, C.; Carroll, L.; Gradkowski, K.; Minzioni, P. Coupling strategies for silicon photonics integrated chips [Invited]. *Photonics Res.* **2019**, *7*, 201–239. [CrossRef]
51. Xu, Y.; Tian, Z.; Meng, X.; Chai, Z. Methods and applications of on-chip beam splitting: A review. *Front. Phys.* **2022**, *10*, 985208. [CrossRef]
52. Han, L.; Kuo, B.P.-P.; Alic, N.; Radic, S. Ultra-broadband multimode 3dB optical power splitter using an adiabatic coupler and a Y-branch. *Opt. Express* **2018**, *26*, 14800–14809. [CrossRef]
53. Krutov, I.A.; Saygin, M.Y.; Dyakonov, I.V.; Kulik, S.P. Optimized low-loss integrated photonics silicon-nitride Y-branch splitter. *AIP Conf. Proc.* **2020**, *2241*, 020027. [CrossRef]
54. Lv, D.; Wu, L.; Liu, C.; Li, A.; Wang, R.; Wu, A. Broadband and Low-Loss Silicon Photonic Directional Coupler for Signal Power Tapping on the 3 µm SOI Waveguide Platform. *Photonics* **2023**, *10*, 776. [CrossRef]
55. Frishman, A.; Malka, D. An Optical 1×4 Power Splitter Based on Silicon–Nitride MMI Using Strip Waveguide Structures. *Nanomaterials* **2023**, *13*, 2077. [CrossRef] [PubMed]
56. Taha, A.M.; Chandran, S.; Viegas, J.; Bian, Y.; Rakowski, M.; Augur, R.; Jacob, A.; Dahlem, M.S. Compact MMI-Based AWGs in a Scalable Monolithic Silicon Photonics Platform. *IEEE Photonics J.* **2021**, *13*, 6600306. [CrossRef]
57. Xu, L.; Chandran, S.; Bian, Y.; Liu, N.; Abumazwed, A.; Lee, W.S.; Zhang, J.; Deptuck, D.; Rakowski, M.; Krishnamoorthy, S.; et al. Low-loss SiN 2 × 2 MMI 50/50 Couplers for Monolithically Integrated Silicon Photonics. In Proceedings of the CLEO 2023, San Jose, CA, USA, 7–12 May 2023; Optica Publishing Group: Washington, DC, USA, 2023; p. JW2A.86. [CrossRef]
58. Vilas, J.; de Cabo, R.F.; Olivares, I.; González-Andrade, D.; Velasco, A.V.; Dias-Ponte, A. Low-Loss Directional Coupler for the C, L and U Bands Based on Subwavelength Gratings. *IEEE Photonics Technol. Lett.* **2023**, *35*, 1331–1334. [CrossRef]
59. Yao, Z.; Wu, K.; Tan, B.X.; Wang, J.; Li, Y.; Zhang, Y.; Poon, A.W. Integrated Silicon Photonic Microresonators: Emerging Technologies. *IEEE J. Sel. Top. Quantum Electron.* **2018**, *24*, 5900324. [CrossRef]
60. Bogaerts, W.; De Heyn, P.P.; Van Vaerenbergh, T.T.; De Vos, K.K.; Selvaraja, S.K.; Claes, T.T.; Dumon, P.; Bienstman, P.; Van Thourhout, D.D.; Baets, R.R. Silicon microring resonators. *Laser Photonics Rev.* **2012**, *6*, 47–73. [CrossRef]
61. Van, V. *Optical Microring Resonators: Theory, Techniques, and Applications*; CRC Press: Boca Raton, FL, USA, 2016. [CrossRef]
62. Nijem, J.; Naiman, A.; Zektzer, R.; Frydendahl, C.; Mazurski, N.; Levy, U. High-Q and high finesse silicon microring resonator. *Opt. Express* **2024**, *32*, 7896–7906. [CrossRef] [PubMed]
63. Xuan, Y.; Liu, Y.; Varghese, L.T.; Metcalf, A.J.; Xue, X.; Wang, P.-H.; Han, K.; Jaramillo-Villegas, J.A.; Al Noman, A.; Wang, C.; et al. High-Q silicon nitride microresonators exhibiting low-power frequency comb initiation. *Optica* **2016**, *3*, 1171–1180. [CrossRef]

64. Cui, S.; Cao, K.; Yu, Y.; Zhang, X. Ultra-high-Q racetrack microring based on silicon-nitride. *arXiv* **2022**, arXiv:2209.01097. [CrossRef]
65. Puckett, M.W.; Liu, K.; Chauhan, N.; Zhao, Q.; Jin, N.; Cheng, H.; Wu, J.; Behunin, R.O.; Rakich, P.T.; Nelson, K.D.; et al. 422 Million intrinsic quality factor planar integrated all-waveguide resonator with sub-MHz linewidth. *Nat. Commun.* **2021**, *12*, 934. [CrossRef]
66. Perestjuk, M.; Armand, R.; Campos, M.G.S.; Ferhat, L.; Reboud, V.; Bresson, N.; Hartmann, J.-M.; Mathieu, V.; Ren, G.; Boes, A.; et al. One million quality factor integrated ring resonators in the mid-infrared. *Nanophotonics* **2025**, *14*, 1009–1015. [CrossRef]
67. Liu, X.; Ruan, Z.; Sun, S.; Guo, C.; Chen, K.; Zhu, N.; Liu, L. Sharp bend and large FSR ring resonator based on the free-form curves on a thin-film lithium niobate platform. *Opt. Express* **2024**, *32*, 9433–9441. [CrossRef]
68. Ouyang, B.; Xing, Y.; Bogaerts, W.; Caro, J. Silicon ring resonators with a free spectral range robust to fabrication variations. *Opt. Express* **2019**, *27*, 38698–38707. [CrossRef]
69. Marques, A.; Cabrita, P.; Carvalhais, M.; Novo, C.; Lima, M.; Rodrigues, F.; Teixeira, A. Simulation and Experimental Validation of a Silicon Photonics Ring Assisted Mach-Zehnder Interferometer Filter. In Proceedings of the 13th International Conference on Photonics, Optics and Laser Technology, Porto, Portugal, 22–24 February 2025; pp. 147–151. [CrossRef]
70. Chen, H.-Z.; Ho, K.-L.; Wang, P.-H. High-Extinction Photonic Filters by Cascaded Mach–Zehnder Interferometer-Coupled Resonators. *Photonics* **2024**, *11*, 1055. [CrossRef]
71. Hiraki, T.; Fukuda, H.; Yamada, K.; Yamamoto, T. Small Sensitivity to Temperature Variations of Si-Photonic Mach–Zehnder Interferometer Using Si and SiN Waveguides. *Front. Mater.* **2015**, *2*, 26. [CrossRef]
72. Yuan, S.; Feng, J.; Yu, Z.; Chen, J.; Liu, H.; Chen, Y.; Guo, S.; Huang, F.; Akimoto, R.; Zeng, H. Silicon Nanowire-Assisted High Uniform Arrayed Waveguide Grating. *Nanomaterials* **2023**, *13*, 182. [CrossRef]
73. Stanton, E.J.; Spott, A.; Davenport, M.L.; Volet, N.; Bowers, J.E. Low-loss arrayed waveguide grating at 760 nm. *Opt. Lett.* **2016**, *41*, 1785–1788. [CrossRef]
74. Stanton, E.J.; Volet, N.; Bowers, J.E. Low-loss arrayed waveguide grating at 2.0 μm. In Proceedings of the Conference on Lasers and Electro-Optics (2017), San Jose, CA, USA, 14–19 May 2017; Optica Publishing Group: Washington, DC, USA, 2017; p. STh1M.7. [CrossRef]
75. Liu, Y.; Wang, X.; Yao, Y.; Du, J.; Song, Q.; Xu, K. Silicon photonic arrayed waveguide grating with 64 channels for the 2 μm spectral range. *Opt. Lett.* **2022**, *47*, 1186–1189. [CrossRef] [PubMed]
76. Siew, S.Y.; Li, B.; Gao, F.; Zheng, H.Y.; Zhang, W.; Guo, P.; Xie, S.W.; Song, A.; Dong, B.; Luo, L.W.; et al. Review of Silicon Photonics Technology and Platform Development. *J. Light. Technol.* **2021**, *39*, 4374–4389. [CrossRef]
77. Lim, A.E.-J.; Song, J.; Fang, Q.; Li, C.; Tu, X.; Duan, N.; Chen, K.K.; Tern, R.P.-C.; Liow, T.-Y. Review of silicon photonics foundry efforts. *IEEE J. Sel. Top. Quantum Electron.* **2013**, *20*, 405–416. [CrossRef]
78. Absil, P.P.; De Heyn, P.; Chen, H.; Verheyen, P.; Lepage, G.; Pantouvaki, M.; De Coster, J.; Khanna, A.; Drissi, Y.; Van Thourhout, D.; et al. Imec iSiPP25G silicon photonics: A robust CMOS-based photonics technology platform. In Proceedings of the Silicon Photonics X, San Francisco, CA, USA, 7–12 February 2015; SPIE: Bellingham, DC, USA, 2015; pp. 166–171. [CrossRef]
79. Ferraro, F.J.; De Heyn, P.; Kim, M.; Rajasekaran, N.; Berciano, M.; Muliuk, G.; Bode, D.; Lepage, G.; Janssen, S.; Magdziak, R.; et al. Imec silicon photonics platforms: Performance overview and roadmap. In Proceedings of the Next-Generation Optical Communication: Components, Sub-Systems, and Systems XII, San Francisco, CA, USA, 31 January–2 February 2023; SPIE: Bellingham, DC, USA, 2023; pp. 22–28. [CrossRef]
80. Quack, N.; Takabayashi, A.Y.; Sattari, H.; Edinger, P.; Jo, G.; Bleiker, S.J.; Errando-Herranz, C.; Gylfason, K.B.; Niklaus, F.; Khan, U.; et al. Integrated silicon photonic MEMS. *Microsyst. Nanoeng.* **2023**, *9*, 27. [CrossRef]
81. Roelkens, G.; Zhang, J.; Bogaert, L.; Billet, M.; Wang, D.; Pan, B.; Kruckel, C.J.; Soltanian, E.; Maes, D.; Vanackere, T.; et al. Micro-Transfer Printing for Heterogeneous Si Photonic Integrated Circuits. *IEEE J. Sel. Top. Quantum Electron.* **2023**, *29*, 8200414. [CrossRef]
82. Sahrling, M.; Li, S.; Lee, F.; Rusu, S. TSMC in the Silicon Photonics Era—An electrical perspective. In Proceedings of the Optical Fiber Communication Conference (OFC) 2025, San Francisco, CA, USA, 30 March–3 April 2025; Optica Publishing Group: Washington, DC, USA, 2025; p. Tu3G.1. [CrossRef]
83. Hsia, H.; Tsai, C.; Ting, K.; Kuo, F.; Lin, C.; Wang, C.; Hou, S.; Chiou, W.; Yu, D.C. Heterogeneous Integration of a Compact Universal Photonic Engine for Silicon Photonics Applications in HPC. In Proceedings of the 2021 IEEE 71st Electronic Components and Technology Conference (ECTC), San Diego, CA, USA, 1 June–4 July 2021; pp. 263–268. [CrossRef]
84. Lu, H.-Y.; Tang, C.-T.; Chen, C.-H.; Tan, B.-X.; Liu, W.-K.; Chiang, H.-Y.; Tsai, S.-Y.; Hsieh, T.-L.; Mao, W.-J.; Lo, W.-T.; et al. Low-loss High-uniformity Silicon Nitride Optical Building Blocks Integrated on Silicon Photonics Platform. In Proceedings of the 2025 Optical Fiber Communications Conference and Exhibition (OFC), San Francisco, CA, USA, 30 March–3 April 2025; pp. 1–3. [CrossRef]
85. Giewont, K.; Hu, S.; Peng, B.; Rakowski, M.; Rauch, S.; Rosenberg, J.C.; Sahin, A.; Stobert, I.; Stricker, A.; Nummy, K.; et al. 300-mm Monolithic Silicon Photonics Foundry Technology. *IEEE J. Sel. Top. Quantum Electron.* **2019**, *25*, 8200611. [CrossRef]

86. Bian, Y.; Hirokawa, T.; Lee, W.S.; Chandran, S.; Giewont, K.; Aboketaf, A.; Liu, Q.; Sporer, R.; Rakowski, M.; Dezfulian, K.; et al. 300-mm monolithic CMOS silicon photonics foundry technology. In Proceedings of the CLEO: Applications and Technology 2024, Charlotte, NC, USA, 5–10 May 2024; p. ATu3H-1.
87. Littlejohns, C.G.; Rowe, D.J.; Du, H.; Li, K.; Zhang, W.; Cao, W.; Bucio, T.D.; Yan, X.; Banakar, M.; Tran, D.; et al. CORNERSTONE's Silicon Photonics Rapid Prototyping Platforms: Current Status and Future Outlook. *Appl. Sci.* **2020**, *10*, 8201. [CrossRef]
88. Tossoun, B.; Xiao, X.; Cheung, S.; Yuan, Y.; Peng, Y.; Srinivasan, S.; Giamougiannis, G.; Huang, Z.; Singaraju, P.; London, Y.; et al. Large-Scale Integrated Photonic Device Platform for Energy-Efficient AI/ML Accelerators. *IEEE J. Sel. Top. Quantum Electron.* **2025**, *31*, 8200326. [CrossRef]
89. Liu, Q.; Aboketaf, A.; Pal, S.; Mosleh, S.; Banihashemian, S.F.; Pavadai, S.; Lee, J.-C.; Bian, Y.; Gong, M.; Orner, B.; et al. High-Power Micro-Ring Modulator and Multi-Channel Coupled Ring Resonator for WDM Design on a 300-mm Monolithic Foundry Platform. In Proceedings of the Optical Fiber Communication Conference (OFC) 2024, San Diego, CA, USA, 24–28 March 2024; Optica Publishing Group: Washington, DC, USA, 2024; p. M3A.2. [CrossRef]
90. Bian, Y.; Aboketaf, A.; Lee, W.S.; Chandran, S.; Huante-Ceron, E.; Najafi, A.; Villafranca, A.; Shiran, H.; Holt, J.; Afzal, F.; et al. Monolithically integrated high-performance Ge-on-Si PIN photodetectors with nearly 60 GHz EO bandwidth, 0.95 A/W re-sponsivity, <30 nA dark current and −36 dB optical return loss enabled by Ge shape optimization. In Proceedings of the Optical Fiber Communication Conference (OFC) 2025, San Francisco, CA, USA, 30 March–3 April 2025; Optica Publishing Group: Washington, DC, USA, 2025; p. Th3E.6. [CrossRef]
91. Bian, Y.; Ramachandran, K.; Wu, Z.-J.; Hedrick, B.; Dezfulian, K.K.; Houghton, T.; Nummy, K.; Fisher, D.; Hirokawa, T.; Donegan, K.; et al. 3D Integrated Laser Attach Technology on a 300-mm Monolithic CMOS Silicon Photonics Platform. *IEEE J. Sel. Top. Quantum Electron.* **2023**, *29*, 8200519. [CrossRef]
92. Xiang, C.; Jin, W.; Huang, D.; Tran, M.A.; Guo, J.; Wan, Y.; Xie, W.; Kurczveil, G.; Netherton, A.M.; Liang, D.; et al. High-Performance Silicon Photonics Using Heterogeneous Integration. *IEEE J. Sel. Top. Quantum Electron.* **2022**, *28*, 8200515. [CrossRef]
93. Xiang, C.; Liu, J.; Guo, J.; Chang, L.; Wang, R.N.; Weng, W.; Peters, J.; Xie, W.; Zhang, Z.; Riemensberger, J.; et al. Laser soliton microcombs heterogeneously integrated on silicon. *Science* **2021**, *373*, 99–103. [CrossRef]
94. Liu, Y.; Qiu, Z.; Ji, X.; Bancora, A.; Lihachev, G.; Riemensberger, J.; Wang, R.N.; Voloshin, A.; Kippenberg, T.J. A fully hybrid integrated erbium-based laser. *Nat. Photonics* **2024**, *18*, 829–835. [CrossRef]
95. Elshaari, A.W.; Pernice, W.; Srinivasan, K.; Benson, O.; Zwiller, V. Hybrid integrated quantum photonic circuits. *Nat. Photonics* **2020**, *14*, 285–298. [CrossRef]
96. Dai, D.; Liu, L.; Gao, S.; Xu, D.-X.; He, S. Polarization management for silicon photonic integrated circuits. *Laser Photonics Rev.* **2013**, *7*, 303–328. [CrossRef]
97. Bian, Y.; Lee, W.S.; Chandran, S.; Hirokawa, T.; Sorbara, M.; Aboketaf, A.; Dezfulian, K.K.; Najafi, A.; Mosleh, S.; Banihashemian, F.; et al. Minimizing polarization dependent loss for monolithic silicon photonic components and receiver circuits. In Proceedings of the CLEO: Applications and Technology 2024, Charlotte, NC, USA, 5–10 May 2024; Optica Publishing Group: Washington, DC, USA, 2024; p. AM2J-5. [CrossRef]
98. Bian, Y.; Lee, W.S.; Chandran, S.; Hirokawa, T.; Sorbara, M.; Aboketaf, A.; Dezfulian, K.K.; Najafi, A.; Mosleh, S.; Banihashemian, F.; et al. Towards polarization insensitive photonic integrated circuits: Polarization dependent loss reduction of CMOSintegrated monolithic SiPh components. In Proceedings of the 2024 Optical Fiber Communications Conference and Exhibition (OFC), San Diego, CA, USA, 24–28 March 2024; IEEE: New York, NY, USA, 2024; pp. 1–3. Available online: https://ieeexplore.ieee.org/abstract/document/10526737/ (accessed on 9 August 2025).
99. Bian, Y.; Ding, H.; Aboketaf, A.; Karra, V.; Sorbara, M.; Hirokawa, T.; Mosleh, S.; Hedges, C.; Lee, W.S.; Chandran, S.; et al. Polarization mode dispersion in CMOS-integrated monolithic SiPh components: Simulations and experiments. In Proceedings of the 2023 Optical Fiber Communications Conference and Exhibition (OFC), San Diego, CA, USA, 5–9 March 2023; pp. 1–3. [CrossRef]
100. Villegas, J.E.; Tamalampudi, S.R.; Rasras, M. High-speed polarization control in integrated silicon photonics. *Opt. Express* **2025**, *33*, 18477–18491. [CrossRef]
101. Shahwar, D.; Yoon, H.H.; Akkanen, S.-T.; Li, D.; Muntaha, S.T.; Cherchi, M.; Aalto, T.; Sun, Z. Polarization management in silicon photonics. *npj Nanophotonics* **2024**, *1*, 35. [CrossRef]
102. Pruessner, M.W.; Tyndall, N.F.; Stievater, T.H. MEMS-tunable polarization management in photonic integrated circuits. *Opt. Express* **2023**, *31*, 31316–31328. [CrossRef]
103. O'Brien, P.; Carrol, L.; Eason, C.; Lee, J.S. Packaging of Silicon Photonic Devices. In *Silicon Photonics III: Systems and Applications*; Pavesi, L., Lockwood, D.J., Eds.; Springer: Berlin/Heidelberg, Germany, 2016; pp. 217–236. [CrossRef]
104. Carroll, L.; Lee, J.-S.; Scarcella, C.; Gradkowski, K.; Duperron, M.; Lu, H.; Zhao, Y.; Eason, C.; Morrissey, P.; Rensing, M.; et al. Photonic Packaging: Transforming Silicon Photonic Integrated Circuits into Photonic Devices. *Appl. Sci.* **2016**, *6*, 426. [CrossRef]
105. Tan, M.; Xu, J.; Liu, S.; Feng, J.; Zhang, H.; Yao, C.; Chen, S.; Guo, H.; Han, G.; Wen, Z.; et al. Co-packaged optics (CPO): Status, challenges, and solutions. *Front. Optoelectron.* **2023**, *16*, 1. [CrossRef] [PubMed]

106. Lee, B.G.; Nedovic, N.; Greer, T.H.; Gray, C.T. Beyond CPO: A Motivation and Approach for Bringing Optics Onto the Silicon Interposer. *J. Light. Technol.* **2023**, *41*, 1152–1162. [CrossRef]
107. Shen, X.; Chen, B.; Zhu, Y.; Shi, W. Silicon photonic integrated circuits and its application in data center. In Proceedings of the Seventh Symposium on Novel Photoelectronic Detection Technology and Applications, Kunming, China, 5–7 November 2020; SPIE: Bellingham, DC, USA, 2021; pp. 2110–2123. [CrossRef]
108. Pira, L.; Antony, A.; Prathap, N.; Peace, D.; Romero, J. Enhanced Photonic Chip Design via Interpretable Machine Learning Techniques. *arXiv* **2025**, arXiv:2505.09266. [CrossRef]
109. Mao, S.; Cheng, L.; Zhao, C.; Khan, F.N.; Li, Q.; Fu, H.Y. Inverse Design for Silicon Photonics: From Iterative Optimization Algorithms to Deep Neural Networks. *Appl. Sci.* **2021**, *11*, 3822. [CrossRef]
110. Shen, R.; Hong, B.; Ren, X.; Yang, F.; Chu, W.; Cai, H.; Huang, W. Recent progress on inverse design for integrated photonic devices: Methodology and applications. *J. Nanophotonics* **2024**, *18*, 010901. [CrossRef]
111. Ma, H.; Huang, J.; Zhang, K.; Yang, J. Inverse-designed arbitrary-input and ultra-compact 1 $\times$ N power splitters based on high symmetric structure. *Sci. Rep.* **2020**, *10*, 11757. [CrossRef]
112. Yang, K.Y.; Shirpurkar, C.; White, A.D.; Zang, J.; Chang, L.; Ashtiani, F.; Guidry, M.A.; Lukin, D.M.; Pericherla, S.V.; Yang, J.; et al. Multi-dimensional data transmission using inverse-designed silicon photonics and microcombs. *Nat. Commun.* **2022**, *13*, 7862. [CrossRef] [PubMed]
113. Rad, H.A.; Ainsworth, T.; Alexander, R.N.; Altieri, B.; Askarani, M.F.; Baby, R.; Banchi, L.; Baragiola, B.Q.; Bourassa, J.E.; Chadwick, R.S.; et al. Scaling and networking a modular photonic quantum computer. *Nature* **2025**, *638*, 912–919. [CrossRef] [PubMed]
114. Alexander, K.; Bahgat, A.; Benyamini, A.; Black, D.; Bonneau, D.; Burgos, S.; Burridge, B.; Cable, H.; Campbell, G.; Catalano, G.; et al. A manufacturable platform for photonic quantum computing. *Nature* **2025**, *641*, 876–883. [CrossRef] [PubMed]
115. Jain, S.; Hlaing, M.H.; Fan, K.-C.; Midkiff, J.; Ning, S.; Feng, C.; Hsiao, P.-Y.; Camp, P.T.; Chen, R.T. Incubating advances in integrated photonics with emerging sensing and computational capabilities. *Appl. Phys. Rev.* **2025**, *12*, 011337. [CrossRef]
116. Sajan, S.C.; Singh, A.; Sharma, P.K.; Kumar, S. Silicon Photonics Biosensors for Cancer Cells Detection—A Review. *IEEE Sens. J.* **2023**, *23*, 3366–3377. [CrossRef]
117. Qiao, Q.; Liu, X.; Ren, Z.; Dong, B.; Xia, J.; Sun, H.; Lee, C.; Zhou, G. MEMS-Enabled On-Chip Computational Mid-Infrared Spectrometer Using Silicon Photonics. *ACS Photonics* **2022**, *9*, 2367–2377. [CrossRef]
118. Zhang, X.; Kwon, K.; Henriksson, J.; Luo, J.; Wu, M.C. A large-scale microelectromechanical-systems-based silicon photonics LiDAR. *Nature* **2022**, *603*, 253–258. [CrossRef] [PubMed]

MDPI AG
Grosspeteranlage 5
4052 Basel
Switzerland
Tel.: +41 61 683 77 34

Photonics Editorial Office
E-mail: photonics@mdpi.com
www.mdpi.com/journal/photonics

www.ingramcontent.com/pod-product-compliance
Lightning Source LLC
Chambersburg PA
CBHW062038220926
51850CB00010B/2590
* 9 7 8 3 7 2 5 8 6 7 4 6 2 *